Lecture Notes in Mathematics

Edited by A. Dold, B. Eckmann and F. Takens

1435

St. Ruscheweyh E.B. Saff
L.C. Salinas R.S. Varga (Eds.)

Computational Methods and Function Theory

Proceedings of a Conference,
held in Valparaíso, Chile, March 13–18, 1989

Springer-Verlag
Berlin Heidelberg New York London
Paris Tokyo Hong Kong Barcelona

Editors

Stephan Ruscheweyh
Mathematisches Institut, Universität Würzburg
8700 Würzburg, FRG

Edward B. Saff
Institute for Constructive Mathematics
Department of Mathematics, University of South Florida
Tampa, Florida 33620, USA

Luis C. Salinas
Departamento de Matemática, Universidad Técnica Federico Santa María
Casilla 110-V, Valparaíso, Chile

Richard S. Varga
Institute for Computational Mathematics, Kent State University
Kent, Ohio 44242, USA

Mathematics Subject Classification (1980): 30B70, 30C10, 30C25, 30C30,
30C70, 30E05, 30E10, 65R20

ISBN 3-540-52768-0 Springer-Verlag Berlin Heidelberg New York
ISBN 0-387-52768-0 Springer-Verlag New York Berlin Heidelberg

Printing and binding: Druckhaus Beltz, Hemsbach/Bergstr.
2146/3140-543210 – Printed on acid-free paper

Preface

This volume contains the proceedings of the international conference on 'Computational Methods and Function Theory', held at the Universidad Técnica Federico Santa María, Valparaíso, Chile, March 13-18, 1989.

That conference had two goals. The first one was to bring together mathematicians representing two somewhat distant areas of research to strengthen the desirable scientific cooperation between their respective disciplines. The second goal was to have this conference in a country where mathematics as a field of research is developing and scientific contacts with foreign experts are very neccessary. It seems that the conference was successful in both regards. Besides, for many of the non-Chilean participants this was the first visit to South-America and these days left them with valuable personal impressions about the regional problems, an experience which may lead to active support and cooperation in the future.

About 40 half- and one-hour lectures were presented during the conference. They are listed on the last pages of this volume. Of course, not all of them led to a contribution for these proceedings since many have been published elsewhere. However, the papers in this volume are fairly representative for the areas covered.

To hold such a conference, in a place somewhat distant from the international mathematical centers, obviously requires strong support from funding agencies, and it is the organizer's pleasure to acknowledge those contributions at this point. The local organization was made possible through generous grants from the Fundación Andes, Chile, and from our host, the Universidad Técnica Federico Santa María. In addition, foreign participants were supported by a special grant of the National Science Foundation (NSF), USA, and by other national agencies such as the Deutsche Forschungsgemeinschaft (DFG), FRG, the German Academic Exchange Service (DAAD), FRG, the British Council, UK, etc.

We also wish to thank the Universidad Técnica Federico Santa María for the hospitality on its marvellous campus overlooking the beautiful Bay of Valparaíso, and the many people who did help us with the organization. Especially, we wish to thank Ruth Ruscheweyh, who assisted the organizers during the conference and the hot phase of its preparation, and also was responsible for the typesetting (in LaTeX) of the papers in this volume. Finally, we should like to thank Springer-Verlag for accepting these proceedings for its Lecture Notes series.

For the editors:
Stephan Ruscheweyh

Contents

Computational Methods and Function Theory
Proceedings, Valparaíso 1989
St. Ruscheweyh, E.B. Saff, L. C. Salinas, R.S. Varga (*eds.*)
Lecture Notes in Mathematics **1435**, pp. 1–26

Open Problems and Conjectures in Complex Analysis

Roger W. Barnard

Department of Mathematics, Texas Tech University
Lubbock, Texas 79409-1042, USA

Introduction

This article surveys some of the open problems and conjectures in complex analysis that the author has been interested in and worked on over the last several years. They include problems on polynomials, geometric function theory, and special functions with a frequent mixture of the three. The problems that will be discussed and the author's collaborators associated with each problem are as follows:

1. Polynomials with nonnegative coefficients

We first discuss a series of conjectures which have as one of their sources the work of Rigler, Trimble and Varga in [66]. In [66] these authors considered two earlier papers by Beauzamy and Enflo [23] and Beauzamy [22], which are connected with polynomials and the classical Jensen inequality. To describe their results, let

$$p(z) = \sum_{j=0}^{m} a_j z^j = \sum_{j=0}^{\infty} a_j z^j, \quad \text{where } a_j = 0,\ j > m,$$

be a complex polynomial ($\not\equiv 0$), let d be a number in the interval $(0, 1)$, and let k be a nonnegative integer. Then (cf [22], [23]) p is said to have concentration d of degree at most k if

$$\sum_{j=0}^{k} |a_j| \geq d \sum_{j=0}^{\infty} |a_j|. \tag{1}$$

Beauzamy and Enflo showed that there exists a constant $\hat{C}_{d,k}$, depending only on d and k, such that for any polynomial p satisfying (1), it is true that

$$\frac{1}{2\pi} \int_0^{2\pi} \log |p(e^{i\theta})| d\theta - \log \left(\sum_{j=0}^{\infty} |a_j| \right) \geq \hat{C}_{d,k}. \tag{2}$$

In the case of $k = 0$ in (2) the inequality is equivalent to the Jensen inequality [23],

$$\frac{1}{2\pi} \int_0^{2\pi} \log |p(e^{i\theta})| d\theta \geq \log |a_0|.$$

Rigler, etc., in [66] considered the extension of this inequality from the class of polynomials to the class of H^∞ (cf. Duren [36]) functions. For $f \in H^\infty$ the functional

$$J(f) := \frac{1}{2\pi} \int_0^{2\pi} \log |f(e^{i\theta})| d\theta - \log \left(\sum_{j=0}^{\infty} |a_j| \right)$$

can be well-defined and is finite. They let

$$C_{d,k} = \inf\{J(f) : f \in H^\infty \text{ and } f(z) = \sum_{j=0}^{\infty} a_j z^j (\not\equiv 0) \text{ satisfies (1)}\}. \tag{3}$$

For a (fixed) $d \in (0, 1)$ and a (fixed) nonnegative integer k, it was shown that there exists an unique positive integer n (dependent on d and k) such that

$$\frac{1}{2^n} \sum_{j=0}^{k} \binom{n}{j} \leq d < \frac{1}{2^{n-1}} \sum_{j=0}^{k} \binom{n-1}{j}.$$

For this n, set

$$\rho = \frac{\binom{n-1}{k}}{\sum_{j=0}^{k}\binom{n-1}{j} - d2^{n-1}} - 1.$$

With these definitions the following conjecture was made in [66].

Conjecture 1. *Let $C_{d,k}$ be defined by (3). Then*

$$C_{d,k} = \log\left(\frac{\rho}{(\rho+1)2^{n-1}}\right). \tag{4}$$

In [66] Conjecture 1 was verified for $k = 0$ and for the subclass of Hurwitz polynomials, i.e., those polynomials with real coefficients and having all their zeros in the left half-plane. In order to verify the conjecture for the entire class an interim step was suggested. This step was one of the motivations for the following problem which was solved recently by this author and others in [10]. Let p be a real polynomial with nonnegative coefficients. Can a conjugate pair of zeros be factored from p so that the resulting polynomial still has nonnegative coefficients? We gave an answer to one proposed choice for factoring out a pair of zeros. Fairly straightforward arguments show that if the degree of the polynomial is less than 6 then a conjugate pair of zeros of *greatest real part* can be factored out and the resulting polynomial will still have non-negative coefficients. However, the example

$$p(z) = 140 + 20z + z^2 + 1000z^3 + 950z^4 + 5z^5 + 20z^6$$

shows that the statement is not true for arbitrary polynomials with non-negative coefficients. A large amount of computer data had suggested the following:

Conjecture 2. *The nonnegativeness of the coefficients of a real polynomial is preserved upon factoring out a conjugate pair of zeros of smallest positive argument in absolute value.*

Interestingly this last conjecture also arose quite independently in the work of Brian Conrey in analytic number theory in his work on one of Polya's conjectures. Conrey announced Conjecture 2 at the annual West Coast Number Theory Conference in December 1987. The conjecture was communicated to this author by the number theorist Ron Evans. Indeed Evans, using a large amount of computer evidence, has generated a closely related conjecture which we include.

Conjecture 3. *If a polynomial of degree $2n$ has zeros*

$$e^{i(t+a_k)} \text{ and } e^{-i(t+a_k)}, \quad k = 1, 2, \cdots n,$$

where the a_k lie between 0 and π, then all the coefficients are nondecreasing functions of t for small $t > 0$ provided the coefficients are all nonnegative for $t = 0$.

A special case of Conjecture 3 where the zeros on the upper semicircle are equally spaced would be of special interest. Although Conjecture 2 was verified in [10] the techniques do not appear applicable to Conjecture 3.

2. The center divided difference of polynomials

Another series of polynomial problems was generated in classical number theory by the work of Evans and Stolarsky in [37]. Given a polynomial p and a real number λ define $\delta_\lambda(p)$, the center divided difference of p, by

$$\delta_\lambda(p) = \begin{cases} \dfrac{p(x+\lambda) - p(z-\lambda)}{2\lambda}, & \lambda \neq 0, \\ p'(z), & \lambda = 0. \end{cases}$$

We did a study of the behavior of the $\delta_\lambda(p)$ as a function of λ in [11]. A number of classical results of Walsh and Obrechkoff and of Kuipers [50] give some information about the zeros of $\delta_\lambda(p)$ as a function of λ. Let $W[p]$ equal the width of the smallest vertical strip containing the zeros of p. It follows from the classical work that

$$W[\delta_\lambda(p)] \leq W[p]$$

and that the diameter of the zero set of $\delta_\lambda(p)$ approaches ∞ as $|\lambda|$ approaches ∞. The Gauss-Lucas theorem shows that

$$W[p'] \leq W[p].$$

It was shown in [11] that

$$W[\delta_\lambda(p)] \leq W[p'] \tag{5}$$

and the conditions on p when equality holds in (5) are given. We were also able to prove that

$$W[\delta_\lambda(p)] = O(1/\lambda) \quad \text{as } |\lambda| \to \infty.$$

The numerical work done by the number theorists had suggested,

Conjecture 4. *$W[\delta_\lambda(p)]$ monotonically decreases to zero as $|\lambda| \to \infty$.*

In that direction it was shown in [11] that

$$W[\delta_{2\lambda}(p)] \leq W[\delta_\lambda(p)] \tag{6}$$

for all positive λ and conditions for equality in (6) were found. In addition, if the zero set of p is symmetric about a vertical line then

$$W[\delta_\lambda(p)] = 0 \text{ for all } \lambda \geq W[p']. \tag{7}$$

However, an example was given of a polynomial p_ε, that contradicts Conjecture 4 at least for some λ. The polynomial p_ε has its zero set symmetric about the imaginary axis and has the property that for small ε, $W[\delta_1(p_\varepsilon)] = 0$ and $W[\delta_\lambda(P_\varepsilon)] = 0$ for $\lambda \geq \sqrt{1+2\varepsilon} = W[p'_\varepsilon]$ while $W[\delta_\lambda(p_\varepsilon)] > 0$ for

$$1 < \lambda < \sqrt{1+2\varepsilon}.$$

Thus conjecture 4 needs to be modified to read

Conjecture 5. *$W[\delta_\lambda(p)]$ monotonically decreases to zero for $\lambda > W[p']$.*

The original question that motivated the number theorist's interest in this problem was the determination of the zeros of $\delta_\lambda(p_N)$ where

$$p_N(z) = \prod_{k=-N}^{N} (z - k).$$

Also occuring in their work were the iterates, $\delta^{(n)}$ of δ defined inductively by

$$\delta_\lambda^{(n)}(p_N) = \delta_\lambda[\delta_\lambda^{(n-1)}(p_N)]$$

with

$$\delta_\lambda^{(1)}(p_N) = \delta_\lambda(p_N).$$

The numerical work had suggested

Conjecture 6. *All nonreal zeros of $\delta_\lambda^{(n)}(p_N)$ are purely imaginary for all λ and all n.*

Conjecture 6 has been verified in [11] for $n = 1$. Indeed, an interesting problem, with other ramifications in number theory, see Stolarsky [71], would be to characterize those polynomials for which $\delta_\lambda^{(n)}$ has only real and pure imaginary roots.

3. Digital filters and zeros of interpolating polynomials

Some interesting problems arise when classical complex analysis techniques are applied to digital filter theory.

Polynomials to be used in interpolation of digital signals are called interpolating polynomials. These polynomials may require modification to assure convergence of their reciprocals on the unit circle. Such modifications provide the opportunity to apply classical analysis theory as was done by the author, Ford, and Wang in [12].

A real function, g, defined for all values of the real independent variable time, t, is called a signal. A digital signal, γ, is a real sequence, $\{\gamma_m : -\infty < m < \infty\}$, consisting of equally spaced values or samples, $\gamma_m = g(m\Delta t)$, from the signal, g, with a time increment or sample interval, Δt. Thus, the independent variable for digital signals such as γ is sample time, $m\Delta t$, or simply sample number, m.

The signal, g, is studied in terms of its classical Fourier transform, G, as a function of real frequency, ω. The digital analog of the Fourier transform consists of the study of a sequence such as γ in terms of its Z-transform, which is defined to be the power series, Γ, having γ_m as the coefficient of z^m. Frequency's digital analog comes from evaluation of Z-transforms such as Γ on the unit circle with the negative of the θ in $z = e^{i\theta}$ referred to as frequency. If the coefficients in Γ are used without any actual evaluation of $\Gamma(z)$ or g is used without computation of G, such use is said to be in the time domain. But if $\Gamma(z)$ is used with evaluation for some z of unit modulus or G is used, such use is said to be in the frequency domain.

Signals are based on even functions in a number of applications. This restricts digital signals to self-inversive cases meaning that $\Gamma(z) = \Gamma(z^{-1})$ for $z \neq 0$. Equivalently, γ is a symmetric sequence meaning that $\gamma_m = \gamma_{-m}$ for all m.

A second signal, f, with Fourier transform, F, poses as a filter of the signal, g, if the convolution integral, $g * f$, of g and f is considered. Of course, the Fourier transform of $g * f$ is the product of the Fourier transforms, G of g and F of f. The discrete analogy consists of the product of Z-transforms, Γ and Φ, where the latter refers to the power series with the sample, $\Phi_m = f(m\Delta t)$, taken from the filter, f, as the coefficient of z^m.

Reduction of certain frequencies is a fundamental aim in the application of a filter, f, to a function, g. This can involve the definition of f by the requirement that $F(\omega)$ be a constant, c, for $|\omega| < \omega_0$ but zero otherwise. If so, c can be chosen so that

$$f(t) = \text{ sinc } \omega_0 t, \tag{8}$$

where sinc is defined by

$$\text{sinc } x = \frac{\sin x}{x}. \tag{9}$$

These equations illustrate that the definition of a real signal is determined from the specifications of its Fourier transform. Similarly, digital signals are often defined by the specification of Z-transforms.

The Fourier transform, F, of the f in (8) is referred to as a frequency window since it has compact support in frequency. Application of such a window to a signal, g, is known as a frequency windowing. These problems concern discrete time windowing. This consists of the scaled truncation of an infinite sequence such as γ to obtain a finite sequence of the form $\{c_m \gamma_m : -L < m < L\}$ wherein the finite sequence, $\{c_m : -L < m < L\}$, is referred to as a time window.

Suppose a given digital signal, $\{b_k : -\infty < k < \infty\}$, is such that b_k is understood to correspond to the time, $kN\Delta t$, with the sample interval, $N\Delta t$, where N is a natural number such that $N > 1$. If this digital signal is to be compared with digital signals based on the smaller sample interval, Δt, the given digital signal must be interpolated to the smaller sample interval, Δt. For example, insertion of $N-1$ zeros between every b_k and b_{k+1}, followed by multiplication of the Z-transform of the result by the interpolating series, P_N, defined by

$$P_N(z) = 1 + \sum_{m=1}^{\infty} (z^m + z^{-m}) \text{ sinc } \frac{m\pi}{N}, \tag{10}$$

leads to

$$A(z) = \sum_{n=-\infty}^{\infty} a_n z^n = \left(\sum_{j=-\infty}^{\infty} b_j z^{jN} \right) P_N(z). \tag{11}$$

Since the coefficient of z^{kN}, a_{kN}, in A comes from products of b_j and $\text{sinc}(m\pi/N)$ such that $kN = jN \pm m$, it follows that $m \equiv 0 \pmod{N}$, $\text{sinc}(m\pi/N)=0$ for nonzero m, and $a_{kN} = b_k$. Thus, A is an interpolation of the given B with coefficients, b_j.

A major goal is to study possible alternatives to the interpolation used in (10) in terms of truncation of the interpolating series in (11). In practice one truncates P to obtain the interpolating polynomial, $P_{N,L}$ defined by

$$P_{N,L}(z) = z^{L-1}\left(1 + \sum_{m=1}^{L-1}(z^m + z^{-m}) \operatorname{sinc}\frac{m\pi}{N}\right), \tag{12}$$

where $N > 1$.

To assure stability and accuracy of evaluation it is important that alternative P's have *no* zeros on the unit circle. It is shown in [12] that all of the zeros of $P_{N,L}$ are of unit modulus when $L \le N$ and examples are given showing that when $L > N + 1$ almost any combination of zeros inside, on, and outside the unit circle can occur. A number of classical results are then combined to give sharp conditions on real sequences $\{c_m : 1 \le m \le \infty\}$ so that the function $P^*_{N,L}$ defined by

$$P^*_{N,L}(z) = z^{L-1}\left[1 + \sum_{m=1}^{L-1}(z^m + z^{-m})c_m \operatorname{sinc}\frac{m\pi}{N}\right] \tag{13}$$

has no zero of unit modulus. In particular, in order to define a useful test to determine if a specific sequence of numbers will work for the c_m's in (13) the following theorem was proved in [12].

Theorem 1. *If a real sequence, $\{b_m : 0 \le m < L, b_0 = 1\}$ is such that*

$$\begin{vmatrix} 1 & b_1 & \cdots & & b_{k-1} & b_k \\ b_1 & 1 & b_1 & \cdots & & b_{k-1} \\ \vdots & & & & & \\ b_{k-1} & \cdots & & b_1 & 1 & b_1 \\ b_k & b_{k-1} & \cdots & & b_1 & 1 \end{vmatrix} \ge 0$$

for $0 < k < L$, let

$$c_m = b_m\left(1 - \frac{2\log L}{L}\right)^m$$

define the coefficients in (13). Then $P^*_{N,L}$ has no zero of unit modulus.

A number of the standard "windows" that occur in the engineering literature are then shown to be just special cases of those defined in Theorem 1, including the very generalized Hamming window and the Hanning window. (see Rabniner and Gold's book, *Theory and Application of Digital Signal Processing.*)

The distribution of zeros and the orthogonality property of the sinc functions determine the interpolating properties in (11) and enables the classical results to be applied. Thus one can ask, can the sinc functions be replaced by more general orthogonal functions, e.g., Jacobi polynomials, to create a more general setting in which many more applications can be found? Discussions with several engineers have suggested this.

4. Omitted values problems

We now discuss a number of open problems in geometric function theory. Let

$$\Delta_r = \{z : |z| < r\}, \text{ with } \Delta_1 = \Delta.$$

Let S denote the class of univalent functions f in Δ normalized by $f(0) = 0$ and $f'(0) = 1$. The problem of omitted values was first posed by Goodman [38] in 1949, restated by MacGregor [57] in his survey article in 1972, then reposed in a more general setting by Brannan [5] in 1977. It also appears in Bernardi's survey article [24] and has appeared in several open problem sets since then including [27],[40] and [60].

For a function f in S, let $A(f)$ denote the Lebesgue measure of the set $\Delta \backslash f(\Delta)$ and let $L(f,r)$ denote the Lebesgue measure of the set $\{\Delta \backslash f(\Delta)\} \bigcap \{w : |w| = r\}$ for some fixed $r, 0 < r < 1$. Two explicit problems posed by Goodman and by Brannan were to determine

$$A = \sup_{F \varepsilon S} A(f), \tag{14}$$

and

$$L(r) = \sup_{F \varepsilon S} L(f,r). \tag{15}$$

Goodman [38] showed that $.22\pi < A < .50\pi$. The lower bound which he obtained was generated by a domain of the type shown in Figure 1.

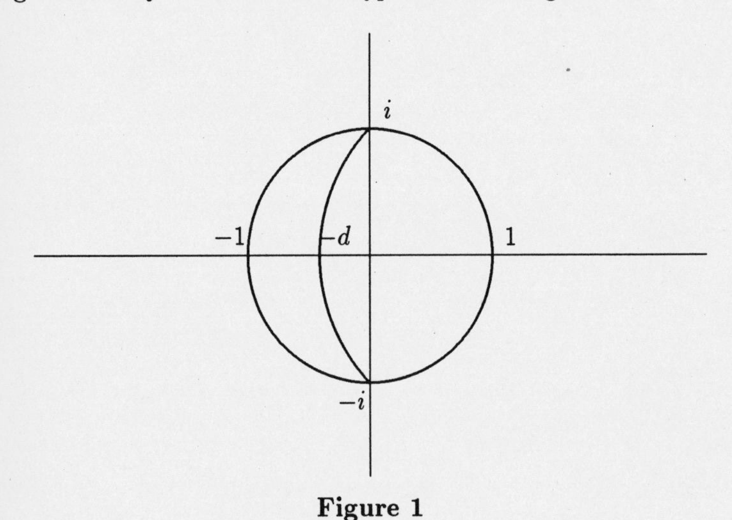

Figure 1

Later, Goodman and Reich [39] gave an improved upper bound of $.38\pi$ for A. Using variational methods developed by the author in [6] and some deep results of Alt and Caffarelli [4] in partial differential equations for free boundary problems, a geometric description for an extremal function for A was given by the author in [9] and by Lewis in [54]. This can be described as follows: There is an f_0 in S with $A = A(f_0)$ such that $f_0(\Delta)$ is circularly symmetric with respect to the positive real axis, i.e., it has the property that for $0 < r < 1$,

$$\frac{\partial}{\partial\theta}|f_0(re^{i\theta})| \text{ and } \frac{\partial}{\partial\theta}|f_0(re^{-i\theta})| \leq 0, \text{ for } 0 < \theta < \pi$$

(cf. Hayman [44]). Moreover the boundary of $f_0(\Delta)$ consists of the negative real axis up to -1, an arc γ of the unit circle that is symmetric about -1 and an arc λ lying in Δ, except for its endpoints. The arc λ is symmetric about the reals, connects the endpoints of γ and has monotonically decreasing modulus in the closure of the upper half disc. These results follow by standard symmetrization methods. Much deeper methods are needed to show (as in [9] and in [54]) that f_0 has a piecewise analytic extension to λ with f_0' continuous on $f_0^{-1}(\lambda)$ and $|f_0'(f_0^{-1}(w))| \equiv c < 1$ for all $w \in \lambda \bigcap \{\Delta \backslash (-1,1)\}$. Using these properties of f_0 it was shown by the author and Pearce in [19] that by "rounding the corners" in certain gearlike domains a close approximation to the extremal function could be obtained. This gives the best known lower bound of

$$.24\pi < A.$$

The upper bound is conceptually harder since it requires an estimate on the omitted area of each function in S. Indeed, it appears difficult to use the geometric description of f_0 to calculate A directly. However, an indirect proof was used by the author and Lewis in [17] to obtain the best known upper bound of

$$A < .31\pi.$$

Open problem. *Show that f_0 is unique and determine A explicitly.*

For the class S^* of functions in S whose images are starlike with respect to the origin, the problem of determining the corresponding

$$A^* = \sup_{f \in S^*} A(f)$$

has been completely solved by Lewis in [54]. The extremal function $f_1 \in S^*$ defined by

$$A^* = A(f_1) \cong .235\pi$$

is unique (up to rotation). The boundary of $f_1(\Delta)$ has two radial rays projecting into Δ with their end points connected by an arc λ_1 that is symmetric about the reals and has $|f_1'(\zeta)| \equiv c_1$ for all $\zeta \in f_1^{-1}(\lambda_1)$.

The problem of determining $L(r)$ in (15) was solved by Jenkins in [47] where he proved that for a fixed $r, 1/4 < r < 1$,

$$L(r) = 2r \arccos(8\sqrt{r} - 8r - 1).$$

The extremal domain in this case is the circular symmetric domain (unique up to rotation) having as its boundary the negative reals up to $-r$ and a single arc of $\{w : |w| = r\}$ symmetric about the point $-r$.

The corresponding problem for starlike functions of determining $L^*(r) = \sup_{f \in S^*} L(f,r)$ was solved by Lewandowski in [53] and by J. Stankiewicz in [70]. The extremal domain in that case is the circularly symmetric domain (unique up to rotation) having as its boundary two radial rays and the single arc of $\{w : |w| = r\}$ connecting their endpoints. An explicit formula for the mapping function in this case was first given by Suffridge in [72].

For the class S^c of functions in S whose images are convex domains the corresponding problem of determining

$$A^c(r) = \sup_{f \in S^c} A(f,r) \tag{16}$$

and

$$L^c(r) = \sup_{f \in S^c} L(r,v). \tag{17}$$

where $A(f,r)$ denotes the Lebesgue measure of $\Delta_r / f(\Delta)$, presents some interesting difficulties. One particular difficulty is that the basic tool of circular symmetrization used in the solution to each of the previous determinations is no longer useful. The example of starting with the convex domain bounded by a square shows that convexity is not always preserved under circular symmetrization. However, Steiner symmetrization (cf. Hayman [44]) can still be used in certain cases such as sectors. Another difficulty is the introduction of distinctly different extremal domains for different ranges of r. Since every function in S^c covers a disk of radius 1/2 (cf. Duren [36]) r needs only to be considered in the interval $(1/2, 1)$. Waniurski has obtained some partial results in [74]. He defined r_1 and r_2 to be the unique solutions to certain transcendental equations where $r_1 \approx .594$ and $r_2 \approx .673$. If $F_{\pi/2}$ is the map of Δ onto the half plane $\{w : \operatorname{Re} w > -1/2\}$ and F_α maps Δ onto the sector

$$\left\{ w : \left| \arg \left(w + \frac{\pi}{4\alpha} \right) \right| < \alpha \right\}$$

whose vertex, $v = -\pi/4\alpha$, is located inside the disk, then

$$\begin{aligned} A^c(r) &= A(F_{\pi/2}, r) \text{ for } 1/2 < r < r_1, \\ L^c(r) &= L(F_{\pi/2}, r) \text{ for } 1/2 < r < r_1, \end{aligned}$$

and

$$L^c(r) = L(F_\alpha, r) \text{ for } r_1 < r < r_2.$$

This author had announced in his survey talk on open problems in complex analysis at the 1985 *Symposium on the Occasion of the Proof of the Bieberbach Conjecture* the following conjecture:

Conjecture 7. *The extremal domains in determining $A^c(r)$ and $L^c(r)$ will be half-planes, symmetric sectors and domains bounded by singles arcs of $|w| = r$ along with tangent lines to the endpoints of these arcs, the different domains depending on different ranges of r in $(1/2, 1)$.*

This conjecture was also made independently by Waniurski at the end of his paper [74] in 1987.

Another conjecture that was announced at the *Symposium on the Proof of the Bieberbach Conjecture* arose out of this author and Pearce's work on the omitted values problem. A significant part of characterizing the extremal domains for $A^c(r)$ and $L^c(r)$ in (16) and (17) via the variational method developed in [6] would be the verification of the following:

Conjecture 8. *If $f \in S^c$ then*

$$\lim_{r\to 1} \frac{1}{2\pi}\int_0^{2\pi}\left|\frac{1}{f'(re^{i\theta})}\right| d\theta \le \sup_{z\in\Delta}\left|\frac{z}{f(z)}\right|. \tag{18}$$

Using standard integral means notation this is equivalent to showing that the smallest c such that

$$\mathcal{M}_1\left[1/f'\right] \le c\mathcal{M}_\infty\left[z/f(z)\right] \tag{19}$$

holds is $c = 1$. Well known results (cf. Duren [36], pp. 214) on integral means show that the smallest c for all functions in S is two, while unpublished results of the author and Pearce show that the smallest c for the class of functions starlike of order $1/2$ [cf Goodman [40]] (a slightly larger class than S^c) is $c = 4/\pi$. It was also shown that equality holds in (18) for all domains bounded by regular polygons and it was conjectured that equality holds for those convex domains bounded by single arcs of $\{w : |w| = r\}$ and tangent lines at the endpoints of these arcs. Verification of Conjecture 8 would give an interesting geometric inequality. Let a convex curve Γ have length L and have its minimum distance from the origin be denoted by d. An application of the isoperimetric inequality along with the conjecture would imply

$$\sqrt{\frac{2d\pi}{L}} \le \sqrt{\frac{1}{2\pi}\int_0^{2\pi}\frac{d\theta}{|f'(e^{i\theta})|}} \le \frac{L}{2d\pi}. \tag{20}$$

We note that the normalization for the functions f in S^c would force the first and last terms in inequality (20) to go to one as d goes to one.

Determining explicit values for $A^c(r)$ and $L^c(r)$ would involve computing the map that takes Δ onto the convex domains bounded by an arc of $\{w : |w| = r\}$ along with the two tangent lines at the endpoints of this arc. The function defining this map involves the quotient of two hypergeometric functions (cf. Nehari, [62]). In particular an extensive verification shows that the function g as shown in Figure 2

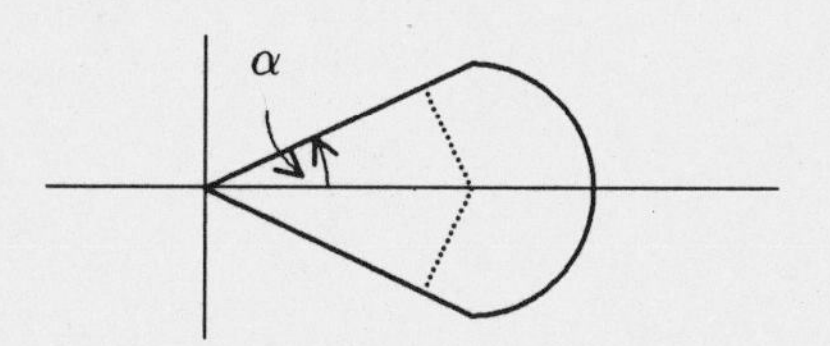

Figure 2

is given by

$$g(z) = \frac{z\,{}_2F_1\left(\frac{2\alpha-1}{4}, \frac{2\alpha+3}{4}, 1+\alpha; z\right)}{{}_2F_1\left(\frac{2\alpha+1}{-4}, \frac{3-2\alpha}{4}, 1-\alpha; z\right)}.$$

A difficulty arises when determining the explicit preimage of the center of the circle so that g can be renormalized to the mapping function f in S taking Δ onto a domain whose boundary circle is centered at the origin.

5. Möbius transformations of convex mappings

Another problem on convex mappings originated from a question of J. Clunie and T. Sheil-Small. If $f \in S$ and $w \notin f(\Delta)$, then the function

$$\hat{f} = f/(1 - f/w) \tag{21}$$

belongs again to S. The transformation $f \to \hat{f}$ is important in the study of geometric function theory. It is useful in the proofs of both elementary and not so elementary properties of S.

If F is a subset of S, let

$$\hat{F} = \{\hat{f} : f \in F, w \in \mathbf{C}^* \backslash f(\Delta)\}.$$

Here $\mathbf{C}^* = \mathbf{C} \bigcup \{\infty\}$. Since we admit $w = \infty$, it is clear that $F \subset \hat{F} \subset S$.

If F is compact in the topology of locally uniform convergence, then so is $\hat{F}$. If F is rotationally invariant, that is, $f_a(z) = e^{ia} f(e^{-ia} z)$ belongs to F whenever f does, then $\hat{F}$ is also rotationally invariant. It is an interesting question to ask which properties of F are inherited by $\hat{F}$. Since $\hat{S} = S$, this question is trivial for S.

In [20] and [21] the author and Schober considered the class S^c of convex mappings. Simple examples show that $\hat{S}^c$ is strictly larger than S^c. Since the coefficients of functions in S^c are uniformly bounded (by one), J. Clunie and T. Sheil-Small had asked whether the coefficients of functions in $\hat{S}^c$ have a uniform bound. The affirmative solution of this problem was given by R.R. Hall [42].

Open Question. Find the best uniform bound as well as the individual coefficient bounds for $\hat{S}^c$.

In [20] the variational procedure developed in [17] is applied to a class of extremal problems for $\hat{S}^c$. If $\lambda : \hat{S}^c \to \mathbf{R}$ is a continuous functional that satisfies certain admissibility criteria, it was shown that the problem

$$\max_{\hat{S}^c} \lambda$$

has a relatively elementary extremal function $\hat{f}$. More specifically, it was shown that $\hat{f}$ either is a half-plane mapping $f(z) = z/(1 - e^{ia} z)$ or is generated through (21) by a parallel strip mapping $f \in S^c$.

The class of functionals considered in [20] contain the second-coefficient functional $\lambda(\hat{f}) = \text{Re } a_2$ and the functionals $\lambda(\hat{f}) = \text{Re } \Phi(\log \hat{f}(z)/z)$ where Φ is entire and z is fixed. The latter functionals include the problems of maximum and minimum modulus ($\Phi(w) = \pm w$). In general, the extremal strip domains $f(\Delta)$ need not be symmetric about the origin. This adds a nontrivial and interesting character to the problems.

A sharp estimate for the second coefficient of functions in $\hat{S}^c$ is given explicitly in the following result. Surprisingly, the answer is not an obvious one.

Theorem 2. *If* $\hat{f}(z) = z + a_2 z^2 + \cdots$ *belongs to* $\hat{S}^c$, *then*

$$|a_2| \leq \frac{2}{x_0} \sin x_0 - \cos x_0 \approx 1.3270$$

where $x_0 \approx 2.0816$ *is the unique solution of the equation*

$$\cot x = \frac{1}{x} - \frac{1}{2} x$$

in the interval $(0, \pi)$. *Equality occurs for the functions* $e^{-ia} \hat{f}(e^{ia} z)$, $\alpha \in \mathbb{R}$, *where* $\hat{f}(z) = f(z)/[1 - f(z)/f(1)]$ *and* f *is the vertical strip mapping defined by*

$$f(z) = \frac{1}{2i \sin x_0} \log \frac{1 + e^{ix_0} z}{1 + e^{-ix_0} z}. \tag{22}$$

We make the following:

Conjecture 9. *The extremal functions for maximizing* $|a_n|$ *over* $\hat{S}^c$ *are the vertical strip mappings defined by (22) where a different* x_0 *is needed for each* n.

In [21] the Koebe disk, radius of convexity, and sharp estimates for the coefficient functional $|ta_3 + a_2^2|$, for t in a certain interval, were found for functions in the class $\hat{S}^c$ Also, in [3], R.M. Ali found sharp upper and lower bounds for $|f(z)|$ for $\hat{f}$ in $\hat{S}^c$.

6. Robinson's 1/2 conjecture

A conjecture that has been open for more than 40 years is Robinson's 1/2 conjecture. Let $\mathcal{A}$ denote the class of analytic functions on Δ. For a subclass X (possibly a singleton) of $\mathcal{A}$ let $r_S(X)$ denote the minimum radius of univalence over all functions f in X.

For a function f in S define the operator $\Theta : S \to \mathcal{A}$ by

$$\Theta f = (zf)'/2.$$

In 1947, in [67], R. Robinson considered the problem of determining $r_S[\Theta(S)]$ which will be denoted by r_0. He observed that for each f in S, $[zf]' \neq 0$ for $\Delta_{1/2}$ and noted that for the Koebe function, $k, k(z) = z(1 - z)^{-2}$,

$$r_S(k) = r_{S^*}(k) = 1/2$$

which implies $r_0 \leq 1/2$. Robinson made

Conjecture 10. *If $f \in S$ then $(zf)'/2$ is univalent in Δ_r for $0 < r \leq 1/2$, i.e. $r_0 = 1/2$.*

He was able to show that $.38 < r_{S^*}[\Theta(S)] \leq r_0$. Little or no progress was made directly on the study of the operator Θ following Robinson's work until Livingston in [56] proposed a shift for the setting of the problem from the full class S to subclasses of S. He showed that Θ preserved many of the well-known subclasses of S. e.g., S^* and S^c. Livingston's work renewed interest in the study of Θ. Numerous papers by various authors followed (see [13]) connecting the operator Θ to various subclasses of S. It was shown by the author and Kellogg in [13] that most of these results follow directly from the Ruscheweyh-Sheil-Small theory on Hadamard convolutions. However, for the entire class S, if appears that the easily obtained lower bound of approximately .41 is the most that can be obtained from the convolution methods. Thus Conjecture 10 is still open. Although Bernardi had suggested that $r_{S^*}[\Theta(S)] = 1/2$ might even be true, in [7], it was shown that $r_{S^*}[\Theta(S)] < .445$, while Pearce proved in [64] that $.435 < r_{S^*}[\Theta(S)]$. In [8] the author proved that $.490 < r_0 \leq .50$ using the Grunsky inequalities. The closeness, but non sharpness, of this result has intrigued a number of people in the field. Robinson's conjecture and the progress on this problem appeared in A. W. Goodman's book, *Univalent Functions* [40], and in [27].

7. Campbell's conjecture on a majorization- subordination result

A conjecture relating majorization and subordination was made by Campbell in [34]. Let f, F, and w be analytic in Δ_r. f is said to be majorized by F, denoted by $f \ll F$, in Δ_r if $|f(z)| \leqq |F(z)|$ in Δ_r. f is said to be subordinate to F, denoted by $f \prec F$, in Δ_r if $f(z) = F(w(z))$ where $|w(z)| \leqq |z|$ in Δ_r.

Majorization-subordination theory began with Biernacki who showed in 1936 that if $f'(0) \geqq 0$ and $f \prec F(F \in S)$ in Δ, then $f \ll F$ in $\Delta_{1/4}$. In the succeeding years Goluzin, Tao Shah, Lewandowski and MacGregor examined various related problems (for greater detail see [33]).

In 1951 Goluzin showed that if $f'(0) \geqq 0$ and $f \prec F(F \in S)$ then $f' \ll F'$ in $\Delta_{0.12}$. He conjectured that majorization would always occur for $|z| < 3 - \sqrt{8}$ and this was proved by Tao Shah in 1958.

In a series of papers [32,33,34], D. Campbell extended a number of the results to the class $\mathcal{U}_\alpha$ of all normalized locally univalent $(f'(z) \neq 0)$ analytic functions in Δ with order $\leqq \alpha$ where $\mathcal{U}_1 = S^c$, the class of convex functions in S. In particular in [34] he showed that if $f'(0) \geqq 0$ and $f \prec F(F \in \mathcal{U}_\alpha)$ then $f' \ll F'$ in $|z| < \alpha + 1 - (\alpha^2 + 2\alpha)^{1/2}$ for $1.65 \leqq \alpha < \infty$ where $\alpha = 2$ yields $3 - \sqrt{8}$. Note that $\alpha = 1$ yields $2 - \sqrt{3}$, the radius of convexity for S. Campbell's proof breaks down for $1 \leqq \alpha < 1.65$ because of two different bounds being used for the Schwarz function with different ranges of α. Nevertheless, he made the following:

Conjecture 11. *If $f'(0) \geq 0$ and $f \prec F$ $(F \in \mathcal{U}_\alpha)$ then $f' \ll F'$ for $|z| < \alpha + 1 - (\alpha^2 + 2\alpha)^{1/2}$.*

In [14] the author and Kellogg combined Ruscheweyh's subordination result [68], variational methods, and some tedious computations to verify the conjecture for $\alpha = 1$, i.e., it is shown that if $f'(0) \geqq 0$ and $f \prec F(F \in S^c)$ in Δ then $f' \ll F'$ for $|z| < 2 - \sqrt{3}$.

8. Krzyż's conjecture for bounded nonvanishing functions

Another conjecture that has been investigated by a large number of function theorists is Krzyż's conjecture. Let B denote the class of functions defined by $f(z) = a_0 + a_1 z + \cdots + a_n z^n + \cdots$ for which $0 < |f(z)| < 1$ for $z \in \Delta$. In 1968 in [49] J. Krzyż posed the fundamental problem of determining for $n \geq 1$

$$A_n = \sup_{f \in B} |a_n|.$$

That $A_1 = 2/e$ dates back to 1932 (see Levin [51]) and appears explicitly in Hummel, etc. [46] and Horowitz [45]. That $A_2 = 2/e$ appears in [46] and $A_3 = 2/e$ in [65]. For a fairly complete history of this problem see [46] or Brown [31]. These results suggest what has become known as the Krzyż Conjecture,

Conjecture 12. *$A_n = 2/e$, for all $n \geq 1$, , with equality only for the functions*

$$K_n(z) = \exp\left[\frac{z^n + 1}{z^n - 1}\right] = \frac{1}{e} + \frac{2}{e} z^n + \cdots$$

and its rotations $e^{iu} K_n(e^{iv} z)$.

A_n is to equal the apocryphal Pondiczery constant, named by Boas in [25]. A sharp uniform bound less than one is expected. However, the bound $2/e \approx .7357\cdots$, is somewhat surprising in view of the fact that the best uniform estimate known to date is

$$|a_n| \leq 1 - \frac{1}{3\pi} + \frac{4}{\pi} \sin\left(\frac{1}{12}\right) = 0.9998772\cdots$$

given by D. Horovitz in 1978 in [45].

The open problem of Krzyż's Conjecture is stated in A. Goodman's book "Univalent Functions" [40, page 83]. De Branges' recent solution to the Bieberbach Conjecture gave hope to solving many of these type problems. However, not withstanding the amount of effort by several function theorists to solve the corresponding coefficient problem, Conjecture 12 still remains open.

A related conjecture made by Ruscheweyh upon verification would give a much improved uniform estimate for A_n. Consider $f(z) = \exp[-\lambda p(z)]$ for $\lambda > 0$ and $p \in P$ where

$$P = \{p : p(z) = 1 + p_1 z + \cdots, \operatorname{Re} p(z) > 0, |z| < 1\}.$$

Then consider the following: For $0 < r < 1$, choose $x = x(r)$ such that

$$i\frac{J_1\left(ix\frac{2r}{1-r^2}\right)}{J_0\left(ix\frac{2r}{1-r^2}\right)} = r \quad (J_0, J_1 \text{ are Bessel Functions})$$

and define

$$F(r) = x(r)e^{-x(r)\frac{1+r^2}{1-r^2}} J_0\left(ix(r)\frac{2r}{1-r^2}\right). \tag{23}$$

Ruscheweyh conjectured that for any positive integer n, $A_k \geq 0, |\xi_k| = 1, k = 1, 2, \cdots,$ n and

$$p(z) = \sum_{k=1}^{n} A_k \left(\frac{1+\zeta_k z}{1-\zeta_k z}\right),$$

Conjecture 13.

$$\frac{1}{2\pi}\int_0^{2\pi} e^{-\mathrm{Re}p(re^{i\varphi})}\mathrm{Re}\{p(re^{i\varphi})\}d\varphi \leq F(r^n), \tag{24}$$

with F defined in (23).

We have shown by using the Legendre polynomial expansion for Bessel functions that

$$\frac{1}{2\pi}\int_0^{2\pi} e^{-x\mathrm{Re}\left\{\frac{1+r^n e^{in\varphi}}{1-r^n e^{in\varphi}}\right\}}\mathrm{Re}\left\{\frac{1+r^n e^{in\varphi}}{1-r^n e^{in\varphi}}\right\}d\varphi = e^{-x\frac{1+r^{2n}}{1-r^{2n}}} J_0\left(ix\frac{2r^n}{1-r^{2n}}\right). \tag{25}$$

Equation (25) shows that the estimate (24) would be sharp for fixed r for $\tilde{p}$ defined by

$$\tilde{p}(z) = x(r^n)\frac{1+z^n}{1-z^n}.$$

Upon verification of Conjecture 13 it can be shown that

$$|a_n| \leq \frac{2}{n}\frac{F(r^n)}{r^{n-1}(1-r^2)}, 0 < r < 1. \tag{26}$$

Choosing $r^2 = (n-1)/(n+1)$ in (26) it would follow that

$$|a_n| \leq \lim_{k\to\infty} \frac{2}{k}\frac{F\left[\left(\frac{k-1}{k+1}\right)^{k/2}\right]}{k\left(\frac{k-1}{k+1}\right)^{\frac{k-1}{2}}\frac{2}{k+1}} = eF\left(\frac{1}{e}\right) \approx .869 \tag{27}$$

by numerical calculations.

9. A conjecture for bounded starlike functions

A conjecture that was made by this author in [6] in 1975 was recently disproved with computer methods by Pearce leaving the problem now as one that probably can only be done numerically. The conjecture involved coefficient estimates for bounded starlike functions in S. Define, for a fixed $M \geq 1$,

$$S_M = \left\{ f \in S : f(z) = z + a_2 z^2 + a_3 z^3 + \cdots, |f(z)| \leq M, z \in \Delta \right\}.$$

The fact that $|a_2|$ is maximized in S_M by the function mapping onto Pick's domain of the disk Δ_M minus a single radial slit has been known since 1917 [see Goodman, vol. I, p.38]. In the early sixties Tammi [73] used Schiffer's variational methods to determine the explicit extremal domains for maximizing the first few coefficients in S_M. In particular he proved that the extremal domains for maximizing $|a_3|$ in S_M are as shown in Figure 3 for the different values of M.

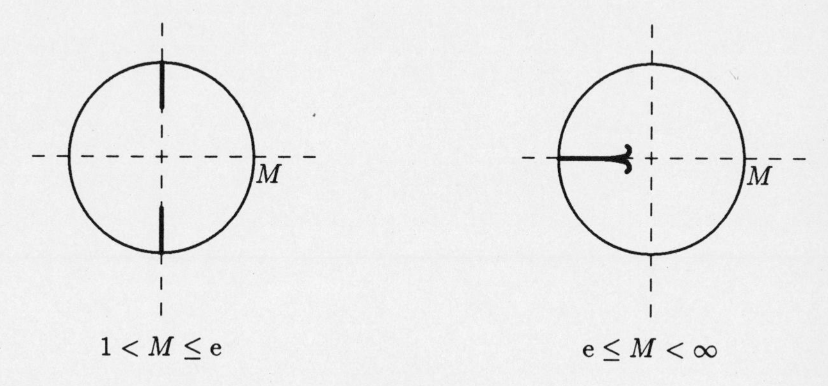

Figure 3

There is a difficulty in modifying Schiffer's variational methods to allow for preservation of both boundedness and starlikeness at the same time. Also the fact that the forked slit domains occurring for $M > 3$ are no longer starlike suggested the need for a local variational technique that preserved these properties. This was developed by combining the Julia Variational formula with the Loewner Theory in [6] and in [17]. Let

$$S_M^* = \{ f \in S_M : f(\Delta) \text{ is starlike with respect to the origin} \}.$$

It was shown in [6] that the extremal domain maximizing $|a_3|$ in S_M^* is the disc Δ_M minus at most two symmetric radial slits. Define D_M as Δ_M minus two symmetric radial slits where 2θ is the angle between the 2 slits. Let $A_3(M,\theta)$ be the third coefficient for the function in S_M^* mapping Δ onto the domain D_M.

From the properties of the extremal domains in the class S_M, along with initial computations and the observation that $A_3(3,0) = A_3(3,\pi/2) = 8/9$ led this author to the following:

Conjecture 14. *For all* $f \in S_M^*$

$$|a_3| \leq A_3(M, \pi/2), \quad 1 < M \leq 3, \tag{28}$$

$$(29) \qquad |a_3| \le A_3(M,0), \quad 3 \le M < \infty.$$

It follows from Tammi's results that (28) holds for $1 < M \le e$ and it was shown by the author and Lewis in [16] that (29) holds for $5 \le M < \infty$. Verifying Conjecture 14 for $e < M < 5$ remained an open problem. This conjecture was announced at the 1978 Brockport Conference and appeared in the open problem set in the proceedings [60] for that conference. It was announced by J. Lewis at the 1980 Canterbury Conference and appeared in the open problem set in its proceedings, [27]. It was also announced at the 1985 *Symposium on the Proof of the Bieberbach Conjecture.*

Motivated by the observation that the domain D_M is indeed a "gearlike" domain and now having the computer software available, Pearce was able to compute $A_3(3,\theta)$ and discovered that $A_3(3,\theta)$, as a function of θ from 0 to $\pi/2$, was convex downward, i.e., it took its *minimum* at the endpoints.

Thus Conjecture 14 was false. Indeed further computations shows that there exists a $\theta(M), 0 < \theta(M) < \pi/2$, such that, for some $M_0 > 0$,

$$\max\left[A_3(M,\theta), A_3(M,\pi/2)\right] < A_3(M,\theta(M))$$

for $2.83 < M < M_0 < 5$.

10. A. Schild's 2/3 conjecture

Another long standing conjecture that was proved false was the 2/3 conjecture. Let $r_1 = r_1(f)$ be the radius of convexity of f, i.e. $r_1(f) = \sup\{r : f(\Delta_r)$ is a convex domain$\}$. Put $d^* = \min\{|f(z)| : |z| = r_1\}$ and $d = \inf|\beta|$ for which $f(z) \neq \beta$. In 1953 in [69], A. Schild conjectured that $d^*/d \ge 2/3$ for all functions $f \in S^*$. Here equality holds for the Koebe function $k(z) = z(1+z)^{-2}$. Schild noted that $d^*/d \ge r_1 \ge 2-\sqrt{3}$ and proved the conjecture for p symmetric functions, $p \ge 7$. He also showed for a certain class of circularly symmetric functions that $d^*/d \ge .49$. Lewandowski in [52], proved the conjecture true for certain subclass of S^*. In [58], McCarty and Tepper obtained the best known lower bound of .38 for all starlike functions. The conjecture was shown false by the author and Lewis in [15] by giving two explicit counter examples.

The first example is given simply by the two slit map defined by

$$f(z) = \frac{z}{(1-z)^{\alpha}(1+z)^{2-\alpha}},$$

where α is sufficiently near 0. It was noted that if d is computed as a function of α, then $\alpha'(d) \to +\infty$ as $\alpha \to 0$ so that a minimal value of .656 for d^*/d was obtained for this function at $\alpha \approx .03$.

The second example is a more complicated function that maps Δ onto the circularly symmetric domain shown in Figure 4.

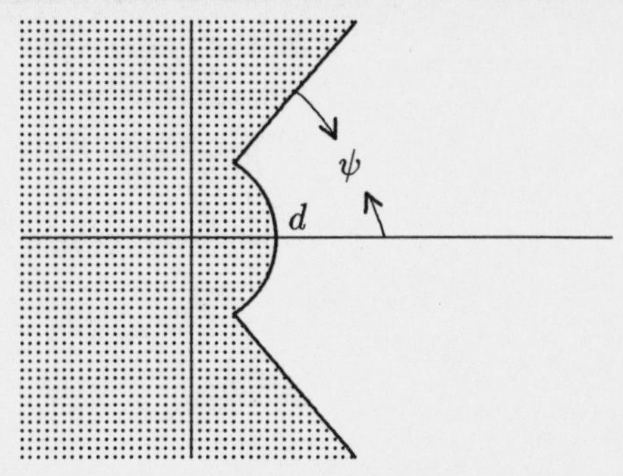

Figure 4

An explicit formula for this function g_a, determined by Suffridge in [72], is given by

$$\log \frac{g_a(z)}{z} = 2b \log \left\{ \left[\left(\frac{1+2az+z^2}{(1-z)^2} \right)^{1/2} + b\frac{1+z}{1-z} \right] \frac{1}{(1+b)} \right\}$$

$$+2\log \frac{2}{\left[(1+2az+z^2)^{1/2}+1+z\right]},$$

where $a = 2b^2 - 1$ and $d = \left[(1+b)^{1+b}(1-b)^{1-b}\right]^{-1}$ with $\psi = \pi(1-b)$. A close approximation to the minimum of d^*/d for this function is 0.644 given by $a \approx 0.89$. Also $\psi \approx .03\pi$ for this minimum value. The author's work suggests:

Conjecture 15.

$$\inf_{f \in S^*} d^*/d = \min_a \{d^*/d \text{ for } g_a\} \approx .644 \cdots$$

11. Brannan's coefficient conjecture for certain power series

An innocent looking, but not so trivial, conjecture was made by Brannan in 1973 in [26] on the coefficients of a specific power series. The problem originated in the Brannan, Clunie, Kirwan paper [28] (later completed by a Aharanov and Friedland in [1]) solving the coefficient problem for functions of bounded boundary rotation. Consider the coefficients in the expansion

$$\frac{(1+xz)^\alpha}{(1-z)^\beta} = \sum_{n=0}^{\infty} A_n^{(\alpha,\beta)}(x) z^n, \quad |x| = 1, \alpha > 0, \beta > 0.$$

Brannan posed the problem as to when

$$\left| A_n^{(\alpha,\beta)}(x) \right| \leq A_n^{(\alpha,\beta)}(1). \tag{30}$$

He gave a short elegant proof that (30) held if $\beta = 1$ and $\alpha \geq 1$. However, he showed that for $\beta = 1, 0 < \alpha < 1$, (30) did *not* hold for the even coefficients and that for $x = e^{i\theta}$, (30) held for odd coefficients in a sufficiently small neighborhood of $\theta = 0$. He also noted that for $0 < \alpha = \beta < 1, |A_3^{(\alpha,\alpha)}(x)| \leq A_3^{(\alpha,\alpha)}(1)$. By using the expansion

$$A_n^{(\alpha,\beta)}(x) = \frac{(\beta)(\beta+1)\cdots(\beta+n-1)}{n!}\ {}_2F_1(-n,\alpha,1-\beta;-x)$$

and the properties of ${}_2F_1$, the hypergeometric function, this author has shown that

(i) (30) holds for $\alpha \leq \beta, \beta + \alpha \geq 1$ and $|A_n^{(\alpha,\beta)}(x)| < A_n^{(\alpha,\beta)}(1)$ for $|x| = 1, x \neq 1$ and $n = 1, 2, 3, \cdots$.

(ii) $|A_{2n+1}^{(\alpha,1)}(x)| \leq A_{2n+1}^{(\alpha,1)}(1), n = 1, 2, 3, \cdots$ for $0 < \alpha < \alpha + \epsilon, 1 - \delta < \alpha < 1$ for ϵ, δ sufficiently small and positive, and

(iii) $|A_3^{(\alpha,\beta)}(x)| \leq A_3^{(\alpha,\beta)}(1), 0 < \alpha < \beta < 1$.

In [61], D. Moak has shown that (30) holds for $\alpha \geq 1, \beta \geq 1$. Milcetich, in [59], has recently shown that (30) holds for $n = 5, \beta = 1$ and $2 < \alpha < n$ but does *not* hold for non integer α's less than $n - 1, \beta$ near zero, for odd $n \geq 3$. The basic

Conjecture 16. $|A_{2n+1}^{(\alpha,1)}(x)| \leq A_{2n+1}^{(\alpha,1)}(1)$

is still open.

12. Polynomial approximations using a differential equation model

Another conjecture on special functions arose out of the author's and L. Reichel's work on polynomial approximations using a differential equation model. Given equidistant data (x_i, y_i) with $x_i = 1 - (2i - 1)/M$, the problem is to *best* fit a polynomial of given degree $N - 1$ to M data points. A comparison is used between the discrete norm $\|\cdot\|_D$, defined by

$$\|f\|_D^2 = \frac{1}{M}\sum_1^M |f(x_i)|^2$$

and the continuous norm, $\|\cdot\|_c$, defined by

$$\|f\|_c = \frac{1}{M}\max_{x\in I}|f(x)|.$$

Gram polynomials $\{\varphi_j\}$ are used where they are orthonormal in the discrete norm with an expansion for p given by

$$p(x) = \sum_j \alpha_j\varphi_j(x),$$

so that $\|p\|_D^2 = \sum \alpha_j^2$. These are defined recursively by

(31) $$\varphi_N(x) = 2x\alpha_{N-1}\varphi_{N-1} - (\alpha_{N-1}/\alpha_{N-2})\,\varphi_{N-2}(x),$$

where

$$\alpha_N = \frac{M}{N}\left(\frac{N^2-1/4}{M^2-N^2}\right)^{1/2}.$$

The asymptotics as M and $N \to \infty$ are studied by letting $\tau = N/\sqrt{M}$ and $x = 1-\zeta/M$. Then the recurrence relation in (31) can be used to obtain

$$\varphi_N - 2\varphi_{N-1} + \varphi_{N-2} = \left[\tau^2 - (1/4\tau^2) - 2\zeta\right]\varphi_N/M + o(1).$$

This in turn can be used to obtain the differential equation model:

(32) $$\varphi''(t) = \left[t^2 - (1/4t^2) - 2\zeta\right]\varphi(t),$$

where $t = \tau - 1/\sqrt{2M}$ and the initial condition as $t \to 0$ is defined by

$$\varphi_N\left[1-\zeta/M\right] = \sqrt{2\sqrt{M}}\sqrt{t} + O(1/M),$$

i.e., $\varphi(t) \approx \sqrt{t}(t \to 0)$. A normalization is made by $\varphi(t) \approx \varphi_n/\sqrt{2\sqrt{M}}$ where ζ is an odd positive integer if and only if x is a grid point. The solution to (32) is given by

$$\varphi(t) = t^{1/2}e^{-t^2/2}\,{}_1F_1\left(\frac{1-\zeta}{2}, 1; t^2\right),$$

where ${}_1F_1$ is Kummer's confluent hypergeometric function (see Gradshteyn and Ryzhik [41]).

To find error estimates for least square approximates by these polynomials an application of Brass's result in [29] can be used that gives error estimates for least square norms in terms of the uniform sup norm. But in order to apply this result all the $\varphi_N(1-\zeta/M)$'s must have their sup norms occur at the right end point of the interval $[-1,1]$. An extensive computer analysis suggested that this does occur. What is needed then is to verify

Conjecture 17. *For all $\zeta > 0$ and real t we have*

$${}_1F_1\left(\frac{1-\zeta}{2}, 1, t^2\right) \le {}_1F_1\left(1/2, 1, t^2\right).$$

Indeed, by converting to the Whittaker functions $M_{\kappa,\mu}(x)$ see [41], for a more convenient range of variables the conjecture is equivalent to showing that

$$M_{\kappa,0}(x) \le M_{0,0}(x) \quad \text{for all } \kappa \ge 0 \text{ and } x \ge 0.$$

We have verified Conjecture 17 for the regions dotted in Figure 5.

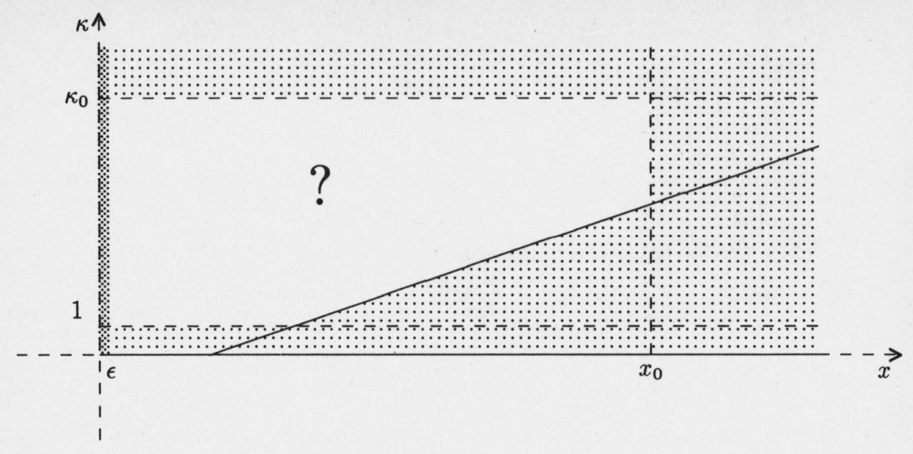

Figure 5

References

[1] D. Aharonov, S. Friedland, *On an inequality connected with the coefficient conjecture for functions of bounded boundary rotation*, Ann. Acad. Sci. Fenn. Ser. A.I. **524** (1972).

[2] H.S. Al-Amiri, *On the radius of univalence of certain analytic functions*, Colloq. Math. **28** (1973) 133-139.

[3] R.M. Ali, *Properties of Convex Mappings*, Ph.D. Thesis, Texas Tech University.

[4] H. Alt, L. Caffarelli, *Existence and regularity for a minimum problem with free boundary*, J. Reine Angew. Math. **325** (1981) 105-144.

[5] J.M. Anderson, K.E. Barth, D.A. Brannan, *Research problems in complex analysis*, Bull. London Math Soc. **9** (1977) 129-162.

[6] R.W. Barnard, *A variational technique for bounded starlike functions*, Canadian Math. J. **27** (1975) 337-347.

[7] R.W. Barnard, *On the radius of starlikeness of $(zf)'$ for f univalent*, Proc. AMS. **53** (1975) 385-390.

[8] R.W. Barnard, *On Robinson's 1/2 conjecture*, Proc. AMS. **72** (1978) 135-139.

[9] R.W. Barnard, *The omitted area problem for univalent functions*, Contemporary Math. (1985) 53-60.

[10] R.W. Barnard, W. Dayawansa, K. Pearce, D. Weinberg, *Polynomials with nonnegative coefficients*, (preprint).

[11] R.W. Barnard, R. Evans, C. FitzGerald, *The center divided difference of polynomials* (preprint).

[12] R.W. Barnard, W.T. Ford, H. Wang, *On the zeros of interpolating polynomials*, SIAM J. of Math. Anal. **17** (1986) 734-744.

[13] R.W. Barnard, C. Kellogg, *Applications of convolution operators to problems in univalent function theory*, Mich. Math. J. **27** (1980) 81-94.

[14] R.W. Barnard, C. Kellogg, *On Campbell's conjecture on the radius of majorization of functions subordinate to convex functions*, Rocky Mountain J. **14** (1984) 331-339.

[15] R.W. Barnard, J.L. Lewis, *A counterexample to the two-thirds conjecture*, Proc. AMS **41** (1973) 525-529.

[16] R.W. Barnard, J.L. Lewis, *Coefficient bounds for some classes of starlike functions.* Pacific J. Math. **56** (1975) 325-331.

[17] R.W. Barnard, J.L. Lewis, *Subordination theorems for some classes of starlike functions*, Pacific J. Math. **56** (1975) 333-366.

[18] R.W. Barnard, Lewis, J.L. *On the omitted area problem*, Mich. Math. J. **34** (1987) 13-22.

[19] R.W. Barnard, K. Pearce, *Rounding corners of gearlike domains and the omitted area problem*, J. Comput. Appl. Math. **14** (1986) 217-226.

[20] R.W. Barnard, G. Schober, *Möbius transformations for convex mappings*, Complex Variables, Theory and Applications **3** (1984) 45-54.

[21] R.W. Barnard, G. Schober, *Möbius transformations for convex mappings II*, Complex Variables, Theory and Applications **7** (1986) 205-214.

[22] B. Beauzamy, *Jensen's inequality for polynomials with concentration at low degrees*, Numer. Math. **49** (1986) 221-225.

[23] B. Beauzamy, P. Enflo, *Estimations de produits de polynômes*, J. Number Theory (1985) **21** 390-412.

[24] S.D. Bernardi, *A survey of the development of the theory of schlicht functions*, Duke Math. J. **19** (1952) 263-287.

[25] R.P. Boas, *Entire Functions* Academic Press, New York, (1954).

[26] D.A. Brannan, *On coefficient problems for certain power series*, Symposium on Complex Analysis, Canterbury, (1973) (ed. Clunie, Hayman). London Math. Soc. Lecture Note Series **12**.

[27] D.A. Brannan, J.G. Clunie, J.G. (eds.). *Aspects of contemporary complex analysis.* (Durham, 1979), Academic Press, London (1980).

[28] D.A. Brannan, J.G. Clunie, W.E. Kirwan, *On the coefficient problem for functions of bounded boundary rotation*, Ann. Acad. Sci. Fenn. Ser. A.I. Math. **3**, (1973).

[29] H. Brass, *Error estimates for least squares approximation by polynomials*, J. Approx. Theory. **41**, (1984) 345-349.

[30] J. Brown, *A coefficient problem for nonvanishing H^p functions*, Rocky Mountain J. of Math. **18** (1988) 707-718.

[31] J. Brown, *A proof of the Krzyż conjecture for* $n = 4$, (preprint).

[32] D. Campbell, *Majorization-Subordination theorems for locally univalent functions*, Bull. AMS **78**, (1972) 535-538.

[33] D. Campbell, *Majorization-Subordination theorems for locally univalent functions II*, Can. J. Math. **25**, (1973) 420-425.

[34] D. Campbell, *Majorization-Subordination theorems for locally univalent functions III*, Trans. Amer. Math. Soc. **198** (1974) 297-306.

[35] S. Chandra, P. Singh, *Certain subclasses of the class of functions regular and univalent in the unit disc*, Arch. Math. **26** (1975) 60-63.

[36] P. Duren, *Univalent Functions.* Springer-Verlag, **259**, New York, (1980).

[37] R.J. Evans, K.B. Stolarsky, *A family of polynomais with concyclic zeros, II*, Proc. AMS. **92**, (1984) 393-396.

[38] A.W. Goodman, *Note on regions omitted by univalent functions*, Bull. Amer. Math. Soc. **55** (1949) 363-369.

[39] A.W. Goodman, E. Reich, *On the regions omitted by univalent functions II*, Canad. J. Math. **7** (1955) 83-88.

[40] A.W. Goodman, *Univalent Functions.* Mariner (1982).

[41] Gradshteyn, Ryzhik. *Table of integral, series and products*, Academic Press, 1980.

[42] R.R. Hall, *On a conjecture of Clunie and Sheil-Small*, Bull, London Math. Soc. **12** (1980) 25-28.

[43] E. Gray, A. Schild, *A new proof of a conjecture of Schild*, Proc. AMS **16** (1965) 76-77, MR30 #2136.

[44] W.K. Hayman, *Multivalent functions*, Cambridge Tracts in Math. and Math. Phys., Cambridge University Press, Cambridge, 1958.

[45] C. Horowitz, *Coefficients of nonvanishing functions in H^∞*, Israel J. Math. **30**, (1978) 285-291.

[46] J.A. Hummel, S. Scheinberg, L. Zalcman, *A coefficient problem for bounded nonvanishing functions*, J. Analyse Math. **31** (1977) 169-190.

[47] J. Jenkins, *On values omitted by univalent functions*, Amer. J. Math. **2** (1953) 406-408.

[48] J. Jenkins, *On circularly symmetric functions*, Proc. AMS. **6** (1955) 620-624.

[49] J. Krzyż, *Coefficient problem for bounded nonvanishing functions*, Ann. Polon. Math. **70** (1968) 314.

[50] L. Kuipers, *Note on the location of zeros of polynomials III*, Simon Stevin. **31** (1957) 61-72.

[51] V. Levin, *Aufgabe 163*, Jber. Dt. Math. Verein. **43** (1933) p. 113, Lösung, ibid, **44** (1934) 80-83 (solutions by W. Fenchel, E. Reissner).

[52] Z. Lewandowski, *Nouvelles remarques sur les théorèmes de Schild relatifs à une classe de fonctions univalentes (Démonstration d'une hypothèse de Schild)*, Ann. Univ. Mariae Curie-Sklodowska Sect. A. **10** (1956).

[53] Z. Lewandowski, *On circular symmetrization of starshaped domains*, Ann. Univ. Mariae Curie-Sklodowska, Sect A. **17** (1963) 35-38.

[54] J.L. Lewis, *On the minimum area problem*, Indiana Univ. Math. J. **34** (1985) 631-661.

[55] R.J. Libera, A.E. Livingston, *On the univalence of some classes of regular functions*, Proc. AMS **30** (1971) 327-336.

[56] A.E. Livingston, *On the radius of univalence of certain analytic functions*, Proc. AMS **17** (1966) 352-357.

[57] T.H. MacGregor, *Geometric problems in complex analysis*, Amer. Math. Monthly. **79** (1972) 447-468.

[58] C. McCarty, D. Tepper, *A note on the 2/3 conjecture for starlike functions*, Proc. AMS **34** (1972) 417-421.

[59] J.G. Milcetich, *On a coefficient conjecture of Brannan*, (preprint).

[60] S. Miller, (ed.). *Complex analysis*, (Brockport, NY, 1976), Lecture Notes in Pure and Appl. Math. **36**, Dekker, New York, (1978).

[61] D. Moak, *An application of hypergeometric function to a problem in function theory*, International J. of Math and Math Sci. **7** (1984).

[62] Z. Nehari, *Conformal Mapping* McGraw Hill, (1952).

[63] K.S. Padmanabhan, *On the radius of univalence of certain classes of analytic functions*, J. London Math. Soc. (2) **1** (1969) 225-231.

[64] K. Pearce, *A note on a problem of Robinson*, Proc. AMS. **89** (1983) 623-627.

[65] D.V. Prokhorov, J. Szynal, *Coefficient estimates for bounded nonvanishing functions*, Bull. Aca. Polon. Sci. Ser. **29** (1981) 223-230.

[66] A.K. Rigler, S.Y. Trimble, R.S. Varga, *Sharp lower bounds for a generalized Jensen Inequality*, Rocky Mountain J. of Math. **19** (1989).

[67] R. Robinson, *Univalent majorants*, Trans. Amer. Math. Soc. **61** (1947) 1-35.

[68] St. Ruscheweyh, *A subordination theorem for Φ-like functions*, J. London Math. Soc. (2) **13** (1973) 275-280.

[69] A. Schild, *On a problem in conformal mapping of schlicht functions*, Proc. AMS **4** (1953) 43-51 MR 14 #861.

[70] J. Stankiewiez, *On a family of starlike functions*, Ann. Univ. Mariae Curie-Sklodowska, Sect A. **22-24** (1968-70) 175-181.

[71] K.B. Stolarsky, *Zeros of exponential polynomials and reductions*, Topics in Classical Number Theory. Collog. Math. Sec., János Bolyai, **34**, Elsevier, (1985).

[72] T. Suffridge, *A coefficient problem for a class of univalent functions*, Mich. Math. J. **16** (1969) 33-42.

[73] O. Tammi, *On the maximalization of the coefficient a_3 of bounded schlicht functions*, Ann. Acad. Sci. Fenn. Ser. AI. **9** (1953).

[74] J. Waniurski, *On values omitted by convex univalent mappings*, Complex Variables, Theory and Appl. **8** (1987) 173-180.

Received: August 30, 1989

Computational Methods and Function Theory
Proceedings, Valparaíso 1989
St. Ruscheweyh, E.B. Saff, L. C. Salinas, R.S. Varga (*eds.*)
Lecture Notes in Mathematics **1435**, pp. 27–31

A Remarkable Cubic Mean Iteration

J.M. Borwein and P.B. Borwein

Mathematics, Statistics and
Computing Science Department
Dalhousie University, Halifax, N.S. B3H 3J5, Canada

1. Introduction

Consider the two term iteration defined by

$$a_{n+1} := \frac{a_n + 2b_n}{3}, \qquad a_0 := a, \tag{1.1}$$

and

$$b_{n+1} := \sqrt[3]{b_n\left(\frac{a_n^2 + a_n b_n + b_n^2}{3}\right)}, \qquad b_0 := b. \tag{1.2}$$

Then since

$$a_{n+1}^3 - b_{n+1}^3 = \frac{(a_n - b_n)^3}{27}, \tag{1.3}$$

it follows that, for $a, b \in (0, \infty)$, and for $n \geq 1$,

$$|a_{n+1} - b_{n+1}| \leq \frac{|a_n - b_n|}{27}$$

and

$$F(a,b) := \lim_{n\to\infty} a_n = \lim_{n\to\infty} b_n \tag{1.4}$$

is well defined, and that on compact subsets of $(0, \infty)$ the convergence is cubic. It is also easy to see that $F(1, z)$ is analytic in some complex neighbourhood of 1. All of this is a straightforward exercise. What is less predictable is that we can identify the limit function explicitly, and that it is a non-algebraic hypergeometric function. Thus, it is one of a very few such examples; and it is certainly the simplest cubic example we know. The most familiar quadratic example is the arithmetic-geometric mean iteration of Gauss and Legendre. Namely the iteration

$$a_{n+1} := \frac{a_n + b_n}{2}, \qquad a_0 := a,$$

$$b_{n+1} := \sqrt{a_n b_n}, \qquad b_0 := x,$$

where, for $0 < x < 1$ $a := 1$

$$\lim_{n\to\infty} a_n = \lim_{n\to\infty} b_n = \frac{1}{{}_2F_1\left(\frac{1}{2},\frac{1}{2};1;1-x^2\right)}.$$

For a discussion of this and a few other examples see [2] and [3].

2. The main theorem

The point of this note is to provide a self-contained proof of the closed form of the limit of (1.1) and (1.2). This is the content of the next theorem.

Theorem 1. *Let $0 < x < 1$. Let*

$$a_{n+1} := \frac{a_n + 2b_n}{3} \qquad a_0 := 1$$

$$b_{n+1} := \sqrt[3]{\frac{b_n(a_n^2 + a_n b_n + b_n^2)}{3}} \qquad b_0 := x.$$

Then the common limit, $F(1,x)$, is

$$\frac{1}{F(1,x)} = \sum_{n=0}^{\infty} \frac{(3n)!}{(n!)^3\, 3^{3n}} \left(1-x^3\right)^n = {}_2F_1\left(\frac{1}{3},\frac{2}{3};1;1-x^3\right).$$

Proof. The limit function $F(a,b)$ must satisfy

$$F(a_0,b_0) = F(a_1,b_1) = \cdots \tag{2.1}$$

and since the iteration is positively homogeneous so is F. In particular

$$F(a_0,b_0) = F(a_1,b_1) = F\left(\frac{a_0+2b_0}{3}, \sqrt[3]{\frac{b_0(a_0^2+a_0b_0+b_0^2)}{3}}\right) \tag{2.2}$$

or

$$\begin{aligned} F(1,x) &= F\left(\frac{1+2x}{3}, \sqrt[3]{\frac{x(1+x+x^2)}{3}}\right) \\ &= \frac{1+2x}{3} F\left(1, \sqrt[3]{\frac{9x(1+x+x^2)}{(1+2x)^3}}\right). \end{aligned} \tag{2.3}$$

If we set $H(x) := \sqrt{x(1-x)}/F(1,(1-x)^{\frac{1}{3}})$ then the functional equation (2.3) becomes

$$H(x) = \sqrt{\frac{3}{t'(x)}}\, H(t(x)), \tag{2.4}$$

where

$$t(x) := 1 - \frac{9x^*(1+x^*+x^{*2})}{(1+2x^*)^3}, \qquad x^* := (1-x)^{\frac{1}{3}}. \tag{2.5}$$

Furthermore $\sqrt{x}H(x)$ is analytic at 0. The point of the proof is to show that

$$G(x) := \sqrt{x(1-x)}\; {}_2F_1\left(\frac{1}{3},\frac{2}{3};1;x\right) \tag{2.6}$$

also satisfies the functional equation (2.4). From this it is easy to deduce that $G(x) = H(x)$; as follows from the functional equation for H/G, and the value at $x = 1$. The (hypergeometric) differential equation satisfied by G is

$$a(x) := \frac{G''(x)}{G(x)} = \left(\frac{-8x^2+8x-9}{36x^2(1-x)^2}\right). \tag{2.7}$$

Now it is a calculation (for details see [2]) that

$$G^*(x) := \sqrt{\frac{3}{t'(x)}} G(t(x)) \tag{2.8}$$

also satisfies (2.7) exactly when

$$a(x) = (t'(x))^2 a(t(x)) - \frac{1}{2}\left[\frac{t'''(x)}{t(x)} - \frac{3}{2}\left(\frac{t''(x)}{t'(x)}\right)^2\right]. \tag{2.9}$$

It is now another calculation, albeit a fairly tedious one, that a and t defined by (2.7) and (2.5) satisfy (2.9). We have now deduced that $G^*(x)$ and $G(x)$ both satisfy (2.7). Furthermore, since the roots of the indicial equation of (2.7) are $(1/2, 1/2)$ there is a fundamental logarithmic solution. Since both G^* and G are asymptotic to $\sqrt{x}$ at 0, they are in fact equal. Thus (2.8) shows that G satisfies (2.4). This finishes the proof. ■

As a consequence we derive the following particularly beautiful cubic hypergeometric transformation.

Corollary 1. *For* $x \in (0,1)$

$${}_2F_1\left(\frac{1}{3},\frac{2}{3};1;1-x^3\right) = \frac{3}{1+2x}\, {}_2F_1\left(\frac{1}{3},\frac{2}{3};1,\left(\frac{1-x}{1+2x}\right)^3\right).$$

Proof. This is just a rewriting of the functional equation (2.3). ■

The above verification entirely obscures our discovery of Theorem 1. This arose from an examination of some quadratic modular equations of Ramanujan [1, Chapter 21]. Notably, Ramanujan observed that,

$$(1-u^3)(1-v^3) = (1-uv)^3 \tag{2.10}$$

is a quadratic modular equation, for ${}_2F_1\left(\frac{1}{3}, \frac{2}{3}; 1; \cdot\right)$. We then observed, with the aid of considerable symbolic computation, that if

$$L(q) := \sum_{-\infty}^{\infty} q^{m^2+mn+n^2} \tag{2.11}$$

and

$$R(q) := \frac{3L(q^3)}{2L(q)} - \frac{1}{2} \tag{2.12}$$

then

$$u := u(q) := R(q) \qquad \text{and} \qquad v := v(q) := R(q^2)$$

solve (2.10) parametrically. From (2.12) it is natural to examine the cubic modular equation for R. This leads to the following result.

Theorem 2. *Let*

$$L(q) := \sum_{-\infty}^{\infty} q^{m^2+mn+n^2}$$

and

$$M(q) := \frac{3L(q^3) - L(q)}{2}.$$

Then, L and M parameterize the mean iteration of (1.1) and (1.2) in the sense that if $a := L(q)$ *and* $b := M(q)$, *then*

$$L(q^3) := \frac{a+2b}{3}$$

and

$$M(q^3) = \sqrt[3]{\frac{b(a^2+ba+b^2)}{3}}$$

and the limit function F (of Theorem 1) satisfies

$$F\left(1, \frac{M(q)}{L(q)}\right) = \frac{1}{L(q)}.$$

The derivation of this, which requires some modular function theory, will be discussed elsewhere [3].

References

[1] B.C. Berndt, *Ramanujan's Notebooks Part II*, Springer-Verlag, New York, 1989.

[2] J.M. Borwein and P.B. Borwein, *Pi and the AGM — A Study in Analytic Number Theory and Computational Complexity*, Wiley N.Y., New York, 1987.

[3] J.M. Borwein and P.B. Borwein, *A Cubic Counterpart of Jacobi's Identity and the AGM*, Trans. A.M.S., to appear.

Received: October 8, 1989

Computational Methods and Function Theory
Proceedings, Valparaíso 1989
St. Ruscheweyh, E.B. Saff, L. C. Salinas, R.S. Varga (*eds.*)
Lecture Notes in Mathematics **1435**, pp. 33–44

On the Maximal Range Problem for Slit Domains[1]

Antonio Córdova Yévenes and Stephan Ruscheweyh
Mathematisches Institut, Universität Würzburg
D-8700 Würzburg, FRG

Abstract. Let $\Omega \subset \mathbb{C}$ be a domain, $0 \in \Omega$. For the family $\mathcal{P}_n(\Omega)$ of complex polynomials p of degree $\leq n$ satisfying $p(0) = 0$, $p(\mathbb{D}) \subset \Omega$ ($\mathbb{D}$ the unit disk) we define the *maximal range* Ω_n as

$$\Omega_n := \bigcup_{p \in \mathcal{P}_n(\Omega)} p(\mathbb{D}).$$

We are interested in the explicit characterization of Ω_n for some specific domains as well as the corresponding *extremal polynomials* $p \in \mathcal{P}_n(\Omega)$, i.e. the ones with $\overline{p(\mathbb{D})} \cap (\partial\Omega_n \setminus \partial\Omega) \neq \emptyset$. In this paper we solve completely the maximal range problem for the slit domains

$$\Omega(a,b) = \mathbb{C} \setminus ((-\infty, -a] \cup [b, \infty)), \qquad a, b > 0.$$

These results yield, for instance, new inequalities relating $\|p\|$, $|\mathrm{Re}\, p|$, $|\mathrm{Im}\, p|$ for typically real polynomials.

1. Introduction

Given a domain Ω in the complex plane $\mathbb{C}$, with $0 \in \Omega$, we define the family $\mathcal{P}_n(\Omega)$ of complex polynomials p of degree at most n which map the unit disk $\mathbb{D}$ into Ω, with the normalized value $p(0) = 0$, i.e.

$$\mathcal{P}_n(\Omega) := \{p \in \mathcal{P}_n : p(0) = 0,\ p(\mathbb{D}) \subset \Omega\}.$$

The *maximal range* for this family is defined as the set

$$\Omega_n := \bigcup_{p \in \mathcal{P}_n(\Omega)} p(\mathbb{D}).$$

[1]Research supported by the Fondo Nacional de Desarrollo Científico y Tecnológico (FONDECYT, Grant 237/89), by the Universidad F. Santa María (Grant 89.12.06), and by the German Academic Exchange Service (DAAD).

We call a polynomial $p \in \mathcal{P}_n(\Omega)$ *extremal* for Ω_n if

$$\overline{p(\mathbb{D})} \cap (\partial\Omega_n \setminus \partial\Omega) \neq \emptyset,$$

and the points $\zeta \in \partial\mathbb{D}$ with $p(\zeta) \in \partial\Omega$ are called the *points of contact* of p.

Our main interest is the description of this set Ω_n and the associated extremal polynomials.

In our previous work on this problem (see [1], [2] and [3]) we gave the explicit characterization of Ω_n and the corresponding extremal polynomials for some typical domains such as interior and exterior of disks, halfplanes and strips. We obtained in this way new sharp estimates relating $\|p\|_{\mathbb{D}}$, $\|\operatorname{Re} p\|_{\mathbb{D}}$, $\|\operatorname{Im} p\|_{\mathbb{D}}$, $\min_{z\in\mathbb{D}} |p(z)|$, etc. for $p \in \mathcal{P}_n(\Omega)$.

The following general characterization for the extremal polynomials, which turned out to be constructive in many cases, has been derived in [3]. For earlier, somewhat weaker versions, see [1], [2].

Theorem 1.1 *Every extremal polynomial $p \in \mathcal{P}_n(\Omega)$ with the normalization $p(1) \in \partial\Omega_n \setminus \partial\Omega$ has the following properties:*

1. *p' has all of its zeros on $\partial\mathbb{D}$. Let $e^{i\psi_j}$, $j = 1, \ldots, n-1$ denote these zeros, where $0 < \psi_1 \leq \cdots \leq \psi_{n-1} < 2\pi$.*
2. *There exist at least n points of contact $e^{i\theta_j}$, $j = 1, \ldots, n$ (multiplicities counted) such that*

$$0 < \theta_1 \leq \psi_1 \leq \theta_2 \leq \psi_2 \leq \cdots \leq \psi_{n-1} \leq \theta_n < 2\pi. \tag{1}$$

3. *If Ω is simply connected, then p is univalent in $\mathbb{D}$.*

Moreover, for every $\omega \in \partial\Omega_n \setminus \partial\Omega$ there exists an extremal polynomial $p \in \mathcal{P}_n(\Omega)$ with $p(1) = \omega$.

In this paper we apply these results to solve completely the maximal range problem for the slit domains

$$\Omega := \mathbb{C} \setminus ((-\infty, -a] \cup [b, +\infty)) \qquad a, b > 0. \tag{2}$$

As an application we obtain a set of new inequalities for polynomials:

Theorem 1.2 *If $P \in \mathcal{P}_n(\Omega)$, $\Omega = \mathbb{C} \setminus [1, \infty)$ then for $z \in \mathbb{D}$*

$$|\operatorname{Im} P(z)| \leq \frac{n+1}{2} \cot \frac{\pi}{n+2},$$

$$-\cot^2 \frac{\pi}{2n+2} \leq \operatorname{Re} P(z) \leq \frac{\cos^2 \frac{\pi}{2n+2}}{\sin \frac{3\pi}{2n+2} \sin \frac{\pi}{2n+2}},$$

$$|P(z)| \leq \cot^2 \frac{\pi}{n+2}.$$

All bounds are best possible.

Theorem 1.3 *Let $\Omega = \mathbb{C} \setminus ((-\infty,-1] \cup [1,\infty))$, n odd. Then for $P \in \mathcal{P}_n(\Omega)$ we have in $\mathbb{D}$*

$$|\mathrm{Re}\, P(z)| \leq \frac{1}{\sin \frac{\pi}{n+1}},$$

$$|P(z)| \leq \frac{n+1}{2}.$$

These bounds are sharp for P as given in (24) at $z = ie^{-\frac{i\pi}{n+1}}$, i, respectively.

Note that this last result holds for typically real polynomials with the normalization $|P(x)| \leq 1, \ -1 < x < 1$.

2. Slit domains

One of the problems which initially inspired us to study the general maximal range problem was to determine the best constants $c(n)$ such that

$$\|p\|_{\mathbb{D}} \leq c(n)\, \|p\|_{[-1,1]} \tag{3}$$

for typically real polynomials p in $\mathbb{D}$ with $p(0) = 0$. Here $\|\cdot\|_{\mathbb{D}}$, $\|\cdot\|_{[-1,1]}$ denote the sup-norm of the modulus on the corresponding sets. This question is related to the work of Rahman and Ruscheweyh [5] and the conjecture was that certain polynomials first studied by Suffridge [4] could be extremal. Clearly, the maximal range problem for the slit domain $\Omega := \mathbb{C} \setminus ((-\infty,-1] \cup [1,+\infty))$ is related to (3) although the condition "typically real" is not referred to when working with maximal ranges in our sense. Fortunately, for n odd, the extremal polynomials turn out to be typically real (the expected ones) thus permitting a solution of (3) in this case. The cases n even remain open for "typically real" (not for the maximal range problem, though).

We study the slightly more general case of the domains $\Omega(a,b)$ as defined in (2). Of course, we can normalize the situation letting $b = 1$ and writing $\Omega(a)$ for $\Omega(a,1)$. We shall show that all extremal polynomials in the sense of Theorem 1.1 for $\Omega(a)$ can be described in terms of the above mentioned polynomials of Suffridge, which are given as follows:

$$P(z;j) = \sum_{k=1}^{n} A_{k,j} z^k, \qquad j = 1, \ldots, n,$$

where

$$A_{k,j} = \frac{n-k+1}{n} \frac{\sin \dfrac{kj\pi}{n+1}}{\sin \dfrac{j\pi}{n+1}}, \quad k, j = 1, \ldots, n.$$

We list a few properties of these polynomials (see [4]). See Figures 1,3 for typical graphs of $P(\partial\mathbb{D}; j)$.

Lemma 2.1 *For $P(z,j)$ the following holds:*

1. *Let $\alpha = \frac{j\pi}{n+1}$. Then, for $|\theta| \le \pi$*

$$P(e^{i\theta};j) = \begin{cases} \dfrac{n+1}{2n(\cos\theta - \cos\alpha)} + i\dfrac{(1-(-1)^j e^{i(n+1)\theta})\sin\theta}{2n(\cos\theta-\cos\alpha)^2}, & \theta \ne \pm\alpha, \\[2ex] \dfrac{(n+1)\cos\alpha}{4n\sin^2\alpha} \pm i\dfrac{(n+1)^2}{4n\sin\alpha}, & \theta = \pm\alpha. \end{cases}$$

2. *$P(z;j)$ is typically real and univalent in $\mathbb{D}$.*

3. *$P'(z;j)$ has all zeros on $\partial\mathbb{D}$.*

4. *$P(e^{i\frac{(2k+j)\pi}{n+1}};j) \in \mathbb{R}$ for $k = 1,\dots,n+1$, $k \ne n+1-j$, $k \ne n+1$.*

5. *$\operatorname{Re} e^{i\frac{(2k+j)\pi}{n+1}} P'(e^{i\frac{(2k+j)\pi}{n+1}};j) = 0, \quad k = 1,\dots,n+1$.*

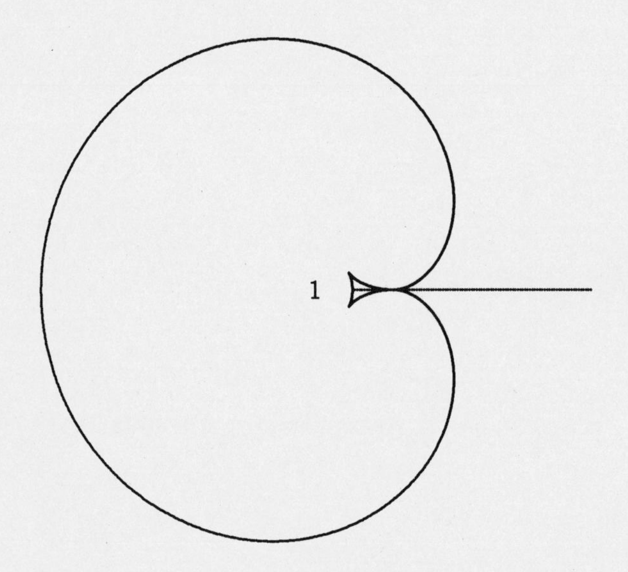

Figure 1

2a. The one-slit case

We start with the one-slit case $\Omega := \Omega(\infty)$. Since Ω is starlike with respect to the origin and symmetric with respect to the real axis the same properties hold true for Ω_n, whose boundary therefore consists of a connected portion of the slit and a Jordan arc connecting the upper and the lower shore of the slit. Let ω be any point on this arc. According to Theorem 1.1 the corresponding extremal polynomial P satisfies

$P \in \mathcal{P}_n(\Omega)$, $\omega \in P(\partial\mathbf{D})$, P is univalent in $\mathbf{D}$, and P' has all the zeros on $\partial\mathbf{D}$, interlacing with points of contact. The velocity of the argument of the tangent vector at $P(e^{i\theta})$ is $\frac{n+1}{2}$, except at the zeros of P' where the argument jumps back by π. Hence between two successive points of contact with the lower (upper) shore θ has to move by exactly $\frac{2\pi}{n+1}$. Near the vertex at 1 there are two possibilities, schematically shown in Fig. 2. In either case, increasing θ by $\frac{2\pi}{n+1}$ from the last point of contact with the lower shore will lead us to the next point of contact ζ with $\operatorname{Re}\zeta p'(\zeta) = 0$ ($p'(\zeta) = 0$ or horizontal tangent). At the other end we find that between the last point of contact with the upper shore and the first one with the lower shore the argument of the tangent vector has to turn by 3π which means for θ a change of $\frac{6\pi}{n+1}$.

Without loss of generality we may assume that the preimage of the last point where P has a horizontal tangent before it reaches the lower shore is $z = 1$. Then we readily obtain the following necessary conditions for P, using $\zeta_k = e^{2\pi ik/(n+1)}$:

$$\operatorname{Im} P(\zeta_k) = 0\ , \quad k = 1, \ldots, n-1, \tag{4}$$

$$\operatorname{Re} \zeta_k P'(\zeta_k) = 0\ , \quad k = 1, \ldots, n+1, \tag{5}$$

$$\sum_{k=1}^{n+1} P(\zeta_k) = 0. \tag{6}$$

The condition (6) follows from the mean value property for polynomials and the assumption $P(0) = 0$. We wish to show that the set of polynomials satisfying (4)–(6) is a one parameter family. However, in the sequel we shall need a slightly more general result.

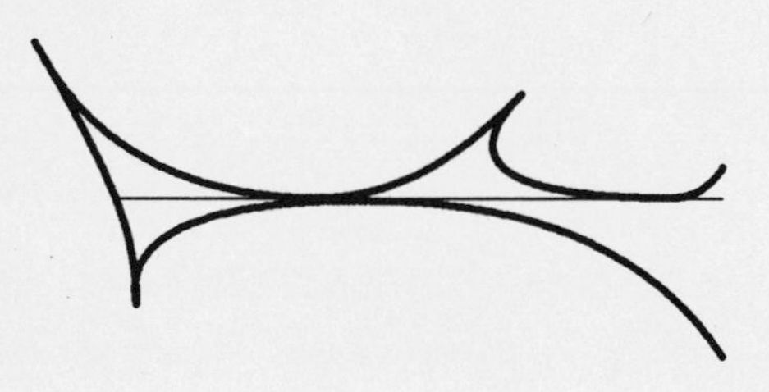

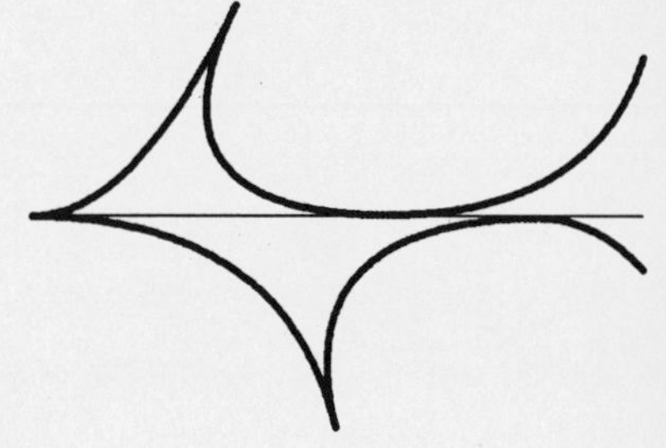

Figure 2

Proposition 2.1 *Let $P \in \mathcal{P}_n$ satisfy (5), (6) and*

$$\operatorname{Im} P(\zeta_k) = 0\ , \qquad k = 1, \ldots, n+1, \quad k \neq r, s, \tag{7}$$

with $r, s \in \{1, \ldots, n+1\}$, $s > r$. Then, for a certain $\lambda \in \mathbf{R}$,

$$P(z) = \lambda P(e^{-i\frac{(s+r)\pi}{n+1}} z; s-r). \tag{8}$$

Furthermore, $\lambda \cdot \operatorname{Re} p(\zeta_k) \geq 0$, $\lambda \cdot \operatorname{Im} p(\zeta_r) \leq 0$.

Proof. Let $Q(z) = 1 - z^{n+1}$. By Lagrange interpolation we obtain

$$P(z) = -\frac{Q(z)}{n+1}\sum_{j+1}^{n+1}\frac{\zeta_j P(\zeta_j)}{z-\zeta_j}$$

and, after differentiation,

$$\zeta_k P'(\zeta_k) = \sum_{\substack{j=1\\ j\neq k}}^{n+1}\frac{\zeta_j P'(\zeta_j)}{\zeta_k-\zeta_j} + \frac{n}{2}P(\zeta_k) \tag{9}$$

$$= \frac{i}{2}\sum_{\substack{j=1\\ j\neq k}}^{n+1} P(\zeta_j)\left[\cot\frac{(j-k)\pi}{n+1}+i\right] + \frac{n}{2}P(\zeta_k). \tag{10}$$

Taking real parts and using (5), (6) we get

$$(n+1)\mathrm{Re}\, P(\zeta_k) = \sum_{\substack{j=1\\ j\neq k}}^{n+1}\mathrm{Im}\, P(\zeta_j)\cot\frac{(j-k)\pi}{n+1}\,, \quad k=1,\dots,n+1. \tag{11}$$

Now let $k = r, s$. Then using (8) we obtain

$$\begin{aligned}(n+1)\mathrm{Re}\, P(\zeta_r) &= \mathrm{Im}\, P(\zeta_s)\cot\frac{(r-s)\pi}{n+1},\\ (n+1)\mathrm{Re}\, P(\zeta_s) &= \mathrm{Im}\, P(\zeta_r)\cot\frac{(s-r)\pi}{n+1}.\end{aligned} \tag{12}$$

Taking imaginary parts in (12) one gets

$$\mathrm{Im}\, P(\zeta_s) = -\mathrm{Im}\, P(\zeta_r), \tag{13}$$

which together with (12) shows that

$$P(\zeta_s) = \overline{P(\zeta_r)}.$$

For $k \neq r, s$ (11) yields

$$\begin{aligned}P(\zeta_k) = \mathrm{Re}\, P(\zeta_k) &= \mathrm{Im}\; P(\zeta_r)\cot\frac{(r-k)\pi}{n+1} + \mathrm{Im}\, P(\zeta_s)\cot\frac{(s-k)\pi}{n+1}\\ &= \mathrm{Im}\, P(\zeta_r)\left[\cot\frac{(r-k)\pi}{n+1} - \cot\frac{(s-k)\pi}{n+1}\right].\end{aligned}$$

Using this and (12) and (13) one deduces that all values $P(\zeta_k)$, $k = 1,\dots,n+1$ are uniquely determined by $\mathrm{Im}\, P(\zeta_r)$ and depend linearly on this parameter. Lemma 2.1 shows that the function on the right hand side of (8) satisfies (5), (6), (7) which implies the representation (8). The final conclusions follow easily. ∎

We return to the one slit case and conditions (4)–(6) for the extremal polynomials. Proposition 2.1 yields (with $r = n$, $s = n + 1$)

$$P(z) = \lambda P(e^{i\frac{\pi}{n+1}}; 1)$$

and by our construction we have $\operatorname{Im} P(\zeta_{n+1}) < 0$, hence $\lambda < 0$. Since rotations are not important for our problem we may use the identity $P(z; j) = P(-z; n + 1 - j)$ to claim that every extremal polynomial (up to rotations) is of the form

$$P(z) = \lambda P(z; n) , \quad \lambda > 0.$$

But $P(z; n)$ has contact with the tip of its corresponding slit at $z = 1$ and this determines λ. We have proved:

Theorem 2.1 *For $\Omega = \mathbb{C} \setminus [1, \infty)$ we have*

$$\partial\Omega_n \setminus \partial\Omega = \left\{ \frac{P(e^{i\theta}; n)}{P(1; n)} : |\theta - \pi| < \frac{3\pi}{n+1} \right\}.$$

Theorem 1.2 is a corollary to this; the bounds for $\operatorname{Im} P$ and $\operatorname{Re} P$ in Theorem 1.2 are attained at $\theta = \frac{n\pi}{n+1}$ and $\theta = \pi$, $\frac{n-1}{n+1}\pi$ respectively, for $P = P(e^{i\theta}; n)$. For estimating $|P(z)|$ we used the fact that the coefficients of $P(-z; n)$ are negative. Figure 1 shows a typical one-slit case ($n = 3$).

2b. The double-slit case

The case of double slits is more complicated. As before, $\Omega_n(a)$ is starlike and symmetric with respect to the real axis. We may assume that the left side has contact with $\Omega_n(a)$ since otherwise we are essentially back at the one slit case. This means that we must take $a \le \cot^2 \frac{\pi}{2n+2}$ (Theorem 1.2). On the other hand, if a is too small, namely $a < \cot^{-2} \frac{\pi}{2n+2}$ then $\partial\Omega_n(a)$ looses contact with the right hand slit and we have again a one slit situation. Hence we may assume that

$$\cot^{-2} \frac{\pi}{2n+2} < a < \cot^2 \frac{\pi}{2n+2}.$$

Then $\partial\Omega_n(a)$ consists of two connected segments of the two slits and two symmetric Jordan curves connecting the upper (lower) shores of the two slits. If ω is a point on the upper Jordan arc then there exists an extremal polynomial P which has contact with both slits and connecting Jordan arc between the slits. The upper arc cannot contain a vertex (a point $P(\zeta)$, with $P'(\zeta) = 0$) but there may be one vertex on the lower one. These observations follow from Theorem 1.1. Again assuming that $P(\zeta_{n+1})$ is the last point with horizontal tangent before we find the first point of contact with the lower shore of the right hand slit we see that for the extremal polynomial

$$\operatorname{Im} P(\zeta_k) = 0 , \quad k = 1, \ldots, n - 1, \quad k \ne r, \tag{14}$$

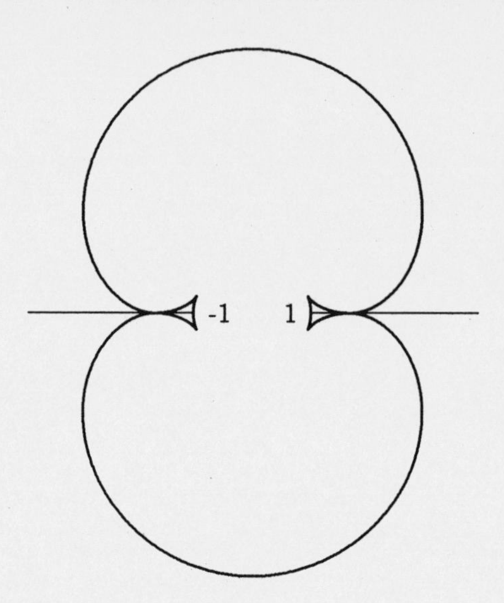

Figure 3

where $r \in \{1, \ldots, n-1\}$. $P(\zeta_r)$ is the point in the upper half plane where P has horizontal tangent. Conditions (5) and (7) remain valid.

We now proceed as in the proof of Proposition 2.1, and arrive at the representation (11). Choosing $k = r, n, n+1$ in (11) we obtain with (14)

$$\begin{aligned}
(n+1)\mathrm{Re}\, P(\zeta_r) &= \mathrm{Im}\, P(\zeta_n) \cot\left(\frac{n-r}{n+1}\pi\right) + \mathrm{Im}\, P(\zeta_{n+1}) \cot\left(\frac{n+1-r}{n+1}\pi\right), \\
(n+1)\mathrm{Re}\, P(\zeta_n) &= \mathrm{Im}\, P(\zeta_r) \cot\left(\frac{r-n}{n+1}\pi\right) + \mathrm{Im}\, P(\zeta_{n+1}) \cot\left(\frac{\pi}{n+1}\right), \\
(n+1)\mathrm{Re}\, P(\zeta_{n+1}) &= \mathrm{Im}\, P(\zeta_r) \cot\left(\frac{r-n-1}{n+1}\pi\right) + \mathrm{Im}\, P(\zeta_n) \cot\left(\frac{-\pi}{n+1}\right).
\end{aligned} \tag{15}$$

and in the remaining cases

$$\begin{aligned}
(n+1)P(\zeta_k) &= (n+1)\mathrm{Re}\, P(\zeta_k) \\
&= \mathrm{Im}\, P(\zeta_r) \cot\left(\frac{r-k}{n+1}\pi\right) + \mathrm{Im}\, P(\zeta_n) \cot\left(\frac{n-k}{n+1}\pi\right) \\
&\quad + \mathrm{Im}\, P(\zeta_{n+1}) \cot\left(\frac{n+1-k}{n+1}\pi\right).
\end{aligned} \tag{16}$$

Finally, taking imaginary parts in (7) we get

$$\mathrm{Im}\, P(\zeta_r) = -\mathrm{Im}\, P(\zeta_n) - \mathrm{Im}\, P(\zeta_{n+1}). \tag{17}$$

Inserting (17) into (15), (16) we deduce that all values $P(\zeta_k)$ depend linearly of the two values $\operatorname{Im} P(\zeta_n)$, $\operatorname{Im} P(\zeta_{n+1})$. Furthermore $\operatorname{Im} P(\zeta_n)$ and $\operatorname{Im} P(\zeta_{n+1})$ are non-positive if we are not in the one slit case which we have excluded. Hence we may write

$$P(z) = \tilde{\alpha} P_1(z) + \tilde{\beta} P_2(z) \ , \quad \tilde{\alpha}, \tilde{\beta} \geq 0,$$

where P_1, P_2 satisfy (14), (5), (6) and $\operatorname{Im} P_1(\zeta_n) = 0$, $P_1(\zeta_{n+1}) < 0$, $\operatorname{Im} P_2(\zeta_n) < 0$, $P_2(\zeta_{n+1}) = 0$. But then we are back at the assumptions of Proposition 2.1 for P_1 (with $s = n+1$) and P_2 (with $s = n$) and we conclude

$$\begin{aligned} P_1(z) &= \lambda_1 P(e^{-\frac{i(n+1+r)}{n+1}\pi} z; n+1-r) \ , \qquad \lambda_1 < 0, \\ P_2(z) &= \lambda_2 P(e^{-\frac{i(n+r)}{n+1}\pi} z; n-r) \ , \qquad \lambda_2 < 0. \end{aligned}$$

Again, using the relation $-P(-z;j) = P(z;n+1-j)$ we can rewrite this as

$$\begin{aligned} P_1(z) &= \tilde{\lambda}_1 P(e^{-\frac{i\pi}{n+1}} \zeta z; r) \ , \qquad \tilde{\lambda}_1 > 0, \\ P_2(z) &= \tilde{\lambda}_2 P(\zeta z; r+1) \ , \qquad \tilde{\lambda}_2 > 0, \end{aligned}$$

with $\zeta = -e^{-i\frac{n+r}{n+1}\pi}$. Since rotations in the argument are of no significance for our problem we finally arrive at

$$P(z) = \alpha P(z;r) + \beta P(e^{i\frac{\pi}{n+1}} z; r+1) \ , \quad \alpha, \beta \geq 0, \tag{18}$$

as a general representation for extremal polynomials related to boundary points of $\Omega_n(a)$ in the upper half plane.

The final step is to show that for every $a \in \left(\cot^{-2} \frac{\pi}{2n+2}, \cot^2 \frac{\pi}{2n+2}\right)$ there is exactly one system (α, β, r) which fulfills the remaining conditions for an extremal polynomial as pointed out in Theorem 1.1. Using (18) and Lemma 2.1, we get for $\theta \neq \pm \frac{r\pi}{n+1}, -\frac{(r+2)\pi}{n+1}$ (mod 2π)

$$\begin{aligned} P(e^{i\theta}) &= \frac{n+1}{2n} [\alpha f_r(\theta) + \beta f_{r+1}(\theta + \gamma)] \\ &= \frac{i}{2n} \left(1 - (-1)^r e^{i(n+1)\theta}\right) \left[\alpha f_r'(\theta) + \beta f_{r+1}'(\theta + \gamma)\right]. \end{aligned} \tag{19}$$

where $\gamma = \frac{\pi}{n+1}$, $f_j(\theta) = 1/(\cos\theta - \cos\frac{j\pi}{n+1})$. Let us now fix r and assume $\alpha, \beta \neq 0$. Then

$$\alpha f_r'(\theta) + \beta f_{r+1}'(\theta + \gamma) \neq 0 \ , \quad \theta = \tfrac{j\pi}{n+1}, \ j = 0, \ldots, n, \ j \neq r$$

and it never changes sign when passing through one of these points. Hence the imaginary part of $P(e^{i\theta})$ cannot change sign in these points as well. But since $P(0) = 0$, the graph of $P(e^{i\theta})$, $\theta \in [0, 2\pi)$ has to cross the real axis at least twice, and this must occur at the zeros of

$$g(\theta) = \alpha f_r'(\theta) + \beta f_{r+1}'(\theta + \gamma).$$

However, the terms of g have different sign only in the intervals $I_1 = (-\gamma, 0)$, $I_2 = (\pi - \gamma, \pi)$. It is easily checked that

$$h(\theta) = \frac{f_r'(\theta)}{f_{r+1}'(\theta+\gamma)} = \frac{\sin\theta}{\sin(\theta+\gamma)}\left(\frac{\cos(\theta+\gamma)-\cos((r+1)\gamma)}{\cos\theta - \cos r\gamma}\right)^2$$

is monotonically increasing from $-\infty$ to 0 in both I_1 and I_2, such that the equation

$$h(\theta) = -\frac{\beta}{\alpha} \tag{20}$$

has exactly two solutions: $\theta_1 \in I_1$ and $\theta_2 \in I_2$. But if P is an extremal polynomial of the type described in Theorem 1.1 then $P(e^{i\theta_1})$, $P(e^{i\theta_2})$ must be the vertices of the two slits otherwise there would be an arc between (and including) the images of two zeros of P' without point of contact (compare Figure 1a). Since $P(e^{i\theta_1}) > 0$, $P(e^{i\theta_2}) < 0$ we can identify the corresponding slits. To have really candidates for our normalized situation $\Omega(a)$ we therefore have to replace P by $P/P(e^{i\theta_1})$. Then, in the new notation, we have

$$P(e^{i\theta_2}) = \frac{\alpha f_r(\theta_2) + \beta f_{r+1}(\theta_2+\gamma)}{\alpha f_r(\theta_1) + \beta f_{r+1}(\theta_1+\gamma)},$$

where we obviously can replace β by $1-\alpha$ and assume $0 < \alpha < 1$. In fact, all considerations remain valid, in a limiting sense, for $\alpha = 1$. Thus all what remains to be done is to show that for each $a \in \left(\cot^{-2}\frac{k}{2n+2}, \cot^2\frac{k}{2n+2}\right)$ there is at most one pair (α, r) with $0 < \alpha < 1$, $r \in \{1, \ldots, n-1\}$ such that

$$-a = \frac{\alpha f_r(\theta_2) + (1-\alpha) f_{r+1}(\theta_2+\gamma)}{\alpha f_r(\theta_1) + (1-\alpha) f_{r+1}(\theta_1+\gamma)}. \tag{21}$$

(that there is at least one follows from Theorem 1.1). If $\alpha = 1$ then $\theta_1 = 0$, $\theta_2 = \pi$ and hence (21) becomes

$$-a = -\tan^2\frac{r\pi}{2n+2} =: -a_r\ , \quad r = 1, \ldots, n-1,$$

a decreasing sequence in r, with $a_n = \cot^2\frac{\pi}{2n+2}$, $a_1 = \cot^{-2}\frac{\pi}{2n+2} = 1/a_n$. Thus we only need to show that the right hand side of (21) is monotone in α for r fixed. This, however is a simple exercise in calculus, differentiating with respect to α and using (20) (with $\beta = 1-\alpha$), whose details we can omit. We arrived at the following theorem:

Theorem 2.2 *Let $a \in [a_r, a_{r+1})$, $r \in \{1, \ldots, n-1\}$. Then there is exactly one $\alpha \in (0,1]$ with*

$$h(\theta_1) = h(\theta_2) = \frac{\alpha - 1}{\alpha}\ , \quad \theta_1 \in (-\tfrac{\pi}{n+1}, 0], \quad \theta_2 \in (\tfrac{n\pi}{n+1}, \pi],$$

for which (21) is satisfied. Then

$$P(z) := \frac{\alpha P(z; r) + (1-\alpha) P(e^{i\frac{\pi}{n+1}} z; r+1)}{\alpha P(e^{i\theta_1}; r) + (1-\alpha) P(e^{i\frac{\pi}{n+1}+\theta_1}; r+1)} \in \mathcal{P}_n(\Omega(a)), \tag{22}$$

and

$$\partial\Omega_n(a) \setminus \partial\Omega(a) = \{P(e^{i\theta}), \overline{P(e^{i\theta})} : \tfrac{r-2}{n+1}\pi < \theta < \tfrac{r+2}{n+1}\pi\}. \tag{23}$$

If $a \geq a_n$ or $a \leq a_1$ then the maximal range problem $\Omega_n(a)$ is essentially a one slit problem and can be solved by Theorem 2.1.

Note that cases $a = a_r$ are simple since then $\alpha = 1$, $\theta_1 = 0$ and the polynomials (22) are typically real. As the special case of probably greatest interest we discuss the one with $a = 1$ (symmetric slit, for an example see Figure 3):

Corollary 2.3 *Let $\Omega = \mathbb{C} \setminus ((-\infty, -1] \cup [1, \infty))$. Then for n odd, we have*

$$P(z) = \frac{P\left(z; \frac{n+1}{2}\right)}{P\left(1; \frac{n+1}{2}\right)} \in \mathcal{P}_n(\Omega), \tag{24}$$

and

$$\partial\Omega_n \setminus \partial\Omega = \{P(e^{i\theta}), \overline{P(e^{i\theta})} : \tfrac{\pi}{2} - \tfrac{2\pi}{n+1} < \theta < \tfrac{\pi}{2} + \tfrac{2\pi}{n+1}\}. \tag{25}$$

For n even, the extremal polynomial and the range Ω_n are described by (22), (23), with $a = 1$, $\alpha = \frac{1}{2}$, $r = [\frac{n+1}{2}]$, and $h(\theta_1) = -1$, $\theta_1 \in (-\frac{\pi}{n+1}, 0)$.

In the case n odd it follows from the properties of the extremal polynomial that $|\mathrm{Re}\, P(e^{i\theta})|$ is maximal for $\theta = \frac{\pi}{2} - \frac{\pi}{n+1}$. The same extremal polynomial is odd and its coefficients are alternating in sign. Hence it takes its maximum modulus at $z = i$. The value at this point can be obtained using Lemma 2.1. This proves

$$|\mathrm{Re}\, P(z)| \leq \frac{1}{\sin \frac{\pi}{n+1}}, \tag{26}$$

$$|P(z)| \leq \frac{n+1}{2}, \tag{27}$$

and Theorem 1.3. We note that Corollary 2.3 holds, in particular, for typically real polynomials in $\mathcal{P}_n(\Omega)$. Hence it contains the solution of the problem (4) for n odd. We obtain the sharp bound

$$c(n) = \frac{n+1}{2}. \tag{28}$$

Clearly, Theorem 1.2 gives also a bound for $c(n)$, n even. However this bound is not sharp since our extremal polynomial is not typically real in this case.

References

[1] Córdova A. and Ruscheweyh St., *On Maximal Ranges of Polynomial Spaces in the Unit Disk*, Constructive Approximation **5** (1989), 309–327.

[2] Córdova A. and Ruscheweyh St., *On Maximal Polynomials Ranges on Circular Domains*, Complex Variables **10** (1988), 295–309.

[3] Córdova A. and Ruscheweyh St., *On the Univalence of Extremal Polynomials for the Maximal Range Problem,* to appear.

[4] Suffridge, T.J., *On Univalent Polynomials* , J. London Math. Soc. **44** (1969), 496-504.

[5] Rahman, Q.I. and Ruscheweyh St., *Markov's Inequality for Typically Real Polynomials*, J. Anal. Appl. (to appear).

Received: August 31, 1989

Computational Methods and Function Theory
Proceedings, Valparaíso 1989
St. Ruscheweyh, E.B. Saff, L. C. Salinas, R.S. Varga (*eds.*)
Lecture Notes in Mathematics **1435**, pp. 45–55

On Bernstein Type Inequalities and a Weighted Chebyshev Approximation Problem on Ellipses [1]

Roland Freund

Institut für Angewandte Mathematik und Statistik, Universität Würzburg,
Am Hubland, D – 8700 Würzburg, FRG
and
RIACS, Mail Stop 230-5, NASA Ames Research Center,
Moffett Field, CA 94035, USA.

Abstract. We are concerned with a classical inequality due to Bernstein which estimates the norm of polynomials on any given ellipse in terms of their norm on any smaller ellipse with the same foci. For the uniform and a certain weighted uniform norm, and for the case that the two ellipses are not "too close", we derive sharp estimates of this type and determine the corresponding extremal polynomials. These Bernstein type inequalities are closely connected with certain constrained Chebyshev approximation problems on ellipses. We also present some new results for a weighted approximation problem of this type.

1. Introduction

Let Π_n denote the set of all complex polynomials of degree at most n. For $r \geq 1$, let

$$\mathcal{E}_r := \{ \, z \in \mathbb{C} \;\mid\; |z-1| + |z+1| \leq r + \frac{1}{r} \, \} \tag{1}$$

be the ellipse with foci at ± 1 and semi-axes $(r \pm 1/r)/2$. Moreover, we use the notation $||\cdot||_{\mathcal{E}_r}$ for the uniform norm $||f||_{\mathcal{E}_r} = \max_{z \in \mathcal{E}_r} |f(z)|$ on $\mathcal{E}_r$.

It is well known (see e.g. [8, Problem III. 271, p. 137]) that, for any $n \in \mathbb{N}$ and $R > r \geq 1$,

$$||p||_{\mathcal{E}_R} \leq \frac{R^n}{r^n} \, ||p||_{\mathcal{E}_r} \quad \text{for all} \quad p \in \Pi_n. \tag{2}$$

---

[1] This work was supported by Cooperative Agreement NCC 2-387 between the National Aeronautics and Space Administration (NASA) and the Universities Space Research Association (USRA).

We remark that for the case $r = 1$, $\mathcal{E}_1 = [-1, 1]$, this inequality goes back to Bernstein (see [2] and the references therein). It is also well known (see [12, p. 368]) that the estimate (2) is not sharp, i.e. equality in (2) holds only for the trivial polynomial $p \equiv 0$.

In this note, we are mainly concerned with the following two problems: find the best possible constants $C_n(r, R)$ and $C_{n+1/2}(r, R)$ such that

$$(3) \qquad ||p||_{\mathcal{E}_R} \leq C_n(r, R)\, ||p||_{\mathcal{E}_r} \quad \text{for all} \quad p \in \Pi_n$$

and

$$(4) \qquad ||wp||_{\mathcal{E}_R} \leq C_{n+1/2}(r, R)\, ||wp||_{\mathcal{E}_r} \quad \text{for all} \quad p \in \Pi_n,$$

respectively. Here, and in the sequel, w denotes the weight function $w(z) = \sqrt{z+1}$, and it is always assumed that the square root is chosen such that w maps the z-plane onto $\{\operatorname{Re} w > 0\} \cup \{i\eta | \eta \geq 0\}$. We notice that the usual proof (e.g. [8, p. 320]) for (2) immediately carries over to the weighted case (4) and leads to the upper bound $C_{n+1/2}(r, R) < R^{n+1/2}/r^{n+1/2}$. For the classical case $r = 1$, Frappier and Rahman [2] conjectured that

$$(5) \qquad C_n(1, R) = \frac{1}{2}(R^n + R^{n-2}) \quad \text{and} \quad C_{n+1/2}(1, R) = \frac{1}{2}(R^{n+1/2} + R^{n-3/2}).$$

The first identity in (5) was proved in [9] for $n = 1$ and in [4] for $n = 2$, $R \geq \sqrt{3}$. The second relation in (5) is known to be true for $n = 1$ and $R \geq 1.49$ [4]. It seems that these are the only cases for which the best possible constants in (3) and (4) are known.

In this paper, sharp estimates (3) and (4) will be obtained for the case $n \in \mathbb{N}$, $r > 1$, and R not "too close" to r. More precisely, we will prove the following

Theorem 1. *Let $n \in \mathbb{N}$ and $r > 1$.*
a) If

$$(6) \qquad R \geq r\, \frac{73r^4 - 1}{r^4 - 1} \quad \textit{resp.} \quad R \geq r\, \frac{33r - 1}{r - 1},$$

then

$$(7) \qquad C_n(r, R) = \frac{R^n + 1/R^n}{r^n + 1/r^n} \quad \textit{resp.} \quad C_{n+1/2}(r, R) = \frac{R^{n+1/2} + 1/R^{n+1/2}}{r^{n+1/2} + 1/r^{n+1/2}}$$

with equality holding in (3) resp. (4) only for the polynomials

$$(8) \quad p(z) \equiv \gamma(T_n(z) + i\frac{2\delta}{R^n - 1/R^n}),\ \gamma \in \mathbb{C},\ \delta \in [-1, 1] \quad \textit{resp.} \quad p(z) \equiv \gamma V_n(z),\ \gamma \in \mathbb{C}.$$

Moreover, if $n = 1$, the first identity in (7) holds true for all $R > r$.
b)

$$C_n(r, R) \geq \frac{R^n + 1/R^n}{r^n + 1/r^n} \quad \textit{and} \quad C_{n+1/2}(r, R) \geq \frac{R^{n+1/2} + 1/R^{n+1/2}}{r^{n+1/2} + 1/r^{n+1/2}}$$

for all $R > r$.

In (8) and in the following, the notation T_k is used for the kth Chebyshev polynomial which by means of the Joukowsky map is given by

$$T_k(z) \equiv \frac{1}{2}(v^k + \frac{1}{v^k}), \quad z \equiv \frac{1}{2}(v + \frac{1}{v}), \tag{9}$$

and, moreover,

$$V_k(z) \equiv \frac{T_{k+1}(z) + T_k(z)}{\sqrt{2}(z+1)}. \tag{10}$$

Notice that (10) defines indeed a polynomial of degree k, and that V_k is up to a scalar factor just the kth Jacobi polynomial $p_k^{(-1/2,1/2)}$ associated with the weight function $(1-z)^{-1/2}(1+z)^{1/2}$ on $[-1,1]$ (cf. Szegö [13, p. 60]).

Remark 1. The estimates (6) are very crude in the following sense. Let $R_n(r)$ resp. $R_{n+1/2}(r)$ denote the smallest numbers such that the first resp. the second identity in (7) is satisfied for all $R \geq R_n(r)$ resp. $R \geq R_{n+1/2}(r)$. Numerical tests reveal that $R_n(r)$ and $R_{n+1/2}(r)$ are much smaller than the upper bounds in (6). Moreover, these experiments suggest that $R_n(r), R_{n+1/2}(r) \approx r$ for large n. However, we were not able to prove these numerical observations.

Although the weighted norms in (4) might appear somewhat artificial, note that (4) arises naturally if, using the Joukowsky map (cf. (9)), one rewrites the estimates (3) and (4) for the disks $|v| \leq R$, $|v| \leq r$, and the class of self-reciprocal polynomials

$$\Sigma_m := \{ \; s \in \Pi_m \mid v^m s(1/v) \equiv s(v) \; \}$$

(cf. [2,4,5]). More precisely,

$$\begin{aligned} & v^n p(\frac{1}{2}(v+\frac{1}{v})) \equiv s(v), \quad p \in \Pi_n, \quad s \in \Sigma_{2n}, \\ \text{resp.} \quad & v^{n+1/2} w(\frac{1}{2}(v+\frac{1}{v})) p(\frac{1}{2}(v+\frac{1}{v})) \equiv s(v), \quad p \in \Pi_n, \quad s \in \Sigma_{2n+1}, \end{aligned} \tag{11}$$

defines a one-to-one mapping between Π_n and Σ_{2n} resp. Σ_{2n+1}. With (11), it is easily verified that (3), (4) are equivalent to

$$\max_{|v| \leq R} |s(v)| \leq D_m(r,R) \max_{|v| \leq r} |s(v)| \quad \text{for all} \quad s \in \Sigma_m, \tag{12}$$

where (3) and (4) correspond to the case $m = 2n$ and $m = 2n+1$, respectively. Moreover, the best possible constants in (3), (4), and (12) are connected by $D_m(r,R) = (R/r)^{m/2} C_{m/2}(r,R)$. Rewriting Theorem 1 for (12), then yields the following

Corollary. *Let $m \geq 2$ be an integer and $r > 1$.*
a) If

$$R \geq r \cdot \begin{cases} (73r^4 - 1)/(r^4 - 1) & \text{if} \quad m \quad \text{is even} \\ (33r - 1)/(r - 1) & \text{if} \quad m \quad \text{is odd} \end{cases},$$

then

$$D_m(r,R)=\frac{R^m+1}{r^m+1}$$

with equality holding in (12) only for the polynomials

$$s(v)\equiv\gamma\,(v^{2n}+i\frac{4\delta}{R^n-1/R^n}\,v^n+1),\quad \gamma\in\mathbb{C},\ \delta\in[-1,1],\quad if\quad m=2n\quad is\ even,$$

and

$$s(v)\equiv\gamma(v^m+1),\quad \gamma\in\mathbb{C},\quad if\quad m\quad is\ odd.$$

Moreover, $D_2(r,R)=(R^2+1)/(r^2+1)$ *for all* $R>r$.
b)

$$D_m(r,R)\geq\frac{R^m+1}{r^m+1}\quad for\ all\quad R>r.$$

Our proof of Theorem 1 is based on the obvious representations

$$C_n(r,R)=\max_{c\in\partial\mathcal{E}_R}\frac{1}{E_n(r,c)}\quad\text{and}\quad C_{n+1/2}(r,R)=\max_{c\in\partial\mathcal{E}_R}\frac{1}{E_{n+1/2}(r,c)}\tag{13}$$

of the sharp constants in (3) and (4) in terms of the optimal values of the family of constrained Chebyshev approximation problems

$$(E_n(r,c):=)\quad\min_{p\in\Pi_n:p(c)=1}\ ||p||_{\mathcal{E}_r}\tag{14}$$

and

$$(E_{n+1/2}(r,c):=)\quad\min_{p\in\Pi_n:p(c)=1}\ ||w_cp||_{\mathcal{E}_r}\quad\text{where}\quad w_c(z)\equiv w(z)/w(c).\tag{15}$$

Here, $c\in\mathbb{C}\setminus\mathcal{E}_r$ in (14) and (15), and, in (13), $\partial\mathcal{E}_R$ denotes the boundary of $\mathcal{E}_R$. The class of complex approximation problems (14) was investigated recently by Fischer and Freund [3]. In particular, the part of Theorem 1 which is concerned with the Bernstein type inequality (3) will follow from results given in [3]. In the present note, we will also derive some new results for the weighted variant (15).

The outline of the paper is as follows. In Section 2, we establish some auxiliary results. In Section 3, the complex weighted approximation problem (15) is studied. Finally, remaining proofs are given in Section 4.

2. Preliminaries

In this section, we introduce some further notations and list two lemmas. It will be convenient to define, in analogy to the Chebyshev polynomials (9), the functions

$$T_{k+1/2}(z)\equiv\frac{1}{2}(v^{k+1/2}+\frac{1}{v^{k+1/2}}),\quad z\equiv\frac{1}{2}(v+\frac{1}{v}),\quad k=0,1,\ldots\ .\tag{16}$$

Here the square root $\sqrt{v}$ is chosen, correspondingly to $w(z) = \sqrt{z+1}$, such that $\sqrt{v}$ maps the v-plane onto $\{\mathrm{Re}\,\xi > 0\} \cup \{i\eta | \eta \geq 1 \text{ or } -1 < \eta \leq 0\}$ (cf. [4]). With (10) and (16), one readily verifies that then

$$T_{k+1/2}(z) = w(z)V_k(z), \quad z \in \mathbb{C}, \tag{17}$$

holds.

For the boundary points $z \in \partial\mathcal{E}_r$ of the ellipse (1), we will use the parametrization

$$z = z_r(\varphi) = \frac{1}{2}(r + \frac{1}{r})\cos\varphi + \frac{i}{2}(r - \frac{1}{r})\sin\varphi, \quad -\pi < \varphi \leq \pi. \tag{18}$$

With (16) and (18), it follows that

$$T_{k+1/2}(z_r(\varphi)) = a_k \cos(k + \frac{1}{2})\varphi + ib_k \sin(k + \frac{1}{2})\varphi, \quad -\pi < \varphi \leq \pi, \tag{19}$$

where

$$a_k := \frac{1}{2}(r^{k+1/2} + \frac{1}{r^{k+1/2}}) \quad \text{and} \quad b_k := \frac{1}{2}(r^{k+1/2} - \frac{1}{r^{k+1/2}}). \tag{20}$$

Next, assume that $n \in \mathbb{N}$ and $r > 1$. Using (17) and (19), we deduce that

$$||wV_n||_{\mathcal{E}_r} = a_n. \tag{21}$$

All corresponding extremal points $z_l \in \mathcal{E}_r$, defined by $|w(z_l)V_n(z_l)| = ||wV_n||_{\mathcal{E}_r}$, are given by

$$z_l := z_r(\varphi_l), \quad \varphi_l := \frac{2l\pi}{2n+1}, \quad l = -n, -n+1, \ldots, n-1, n. \tag{22}$$

Moreover, we note that, in view of (17), (19), and (22),

$$V_n(z_l) = (-1)^l \frac{a_n}{w(z_l)}, \quad l = -n, \ldots, n. \tag{23}$$

The following property of the numbers φ_l will be used in the next section.

Lemma 1. *Let $j \in \mathbb{Z}$. Then:*

$$\sum_{l=-n}^{n} (-1)^l e^{i\varphi_l(j+\frac{1}{2})} = \begin{cases} 2n+1 & \text{if } \frac{2j+1}{2n+1} \in 2\mathbb{Z}+1 \\ 0 & \text{otherwise} \end{cases}. \tag{24}$$

Proof. With $q := (2j+1)/(2n+1)$, we have $e^{i\varphi_l(j+1/2)} = (e^{q\pi i})^l$. If $q \in 2\mathbb{Z}+1$, then $e^{q\pi i} = -1$, and (24) is obviously true in this case. For $q \notin 2\mathbb{Z}+1$, (24) follows from

$$\sum_{l=-n}^{n} (-e^{q\pi i})^l = (-e^{q\pi i})^{-n} \frac{1 - (-e^{q\pi i})^{2n+1}}{1 + e^{q\pi i}}$$

and $(-e^{q\pi i})^{2n+1} = 1$. ∎

Finally, we will apply the following result due to Rogosinški and Szegö [11] in Section 4.

Lemma 2. *Let $\lambda_0, \lambda_1, \ldots, \lambda_n$ be real numbers which satisfy $\lambda_n \geq 0$, $\lambda_{n-1} - 2\lambda_n \geq 0$, and $\lambda_{k-1} - 2\lambda_k + \lambda_{k+1} \geq 0$ for $k = 1, 2, \ldots, n-1$. Then:*

$$t(\varphi) := \frac{\lambda_0}{2} + \sum_{k=1}^{n} \lambda_k \cos(k\varphi) \geq 0 \quad \text{for all} \quad \varphi \in \mathbb{R}. \tag{25}$$

3. Results for the weighted approximation problem (15)

In this section, we are concerned with the constrained Chebyshev approximation problem (15). In the sequel, it is assumed that $n \in \mathbb{N}$, $r \geq 1$, and $c \in \mathbb{C} \setminus \mathcal{E}_r$. Standard results from approximation theory (see e.g. [6]) then guarantee that there always exists a unique optimal polynomial for (15). For the case $r = 1$ of the unit interval $\mathcal{E}_1 = [-1, 1]$ and $c \in \mathbb{R} \setminus [-1, 1]$, Bernstein [1] proved that the rescaled polynomial (10)

$$v_n(z; c) \equiv \frac{V_n(z)}{V_n(c)} \tag{26}$$

is the extremal function for (15). For purely imaginary c and, again, $r = 1$, Freund and Ruscheweyh [4] showed that the optimal polynomial is a suitable linear combination of v_n, v_{n-1}, and v_{n-2}. To the best of our knowledge, these two cases seem to be the only ones for which the solution of (15) is explicitly known.

For the rest of the paper, we assume that $r > 1$. It turns out that, somewhat surprisingly, (26) is also best possible for the general class (15) with complex c as long as c is not "too close" to $\mathcal{E}_r$. For the following, it will be convenient, to represent $c \notin \mathcal{E}_r$ in the form

$$c = c_r(\psi) = \frac{1}{2}(R + \frac{1}{R}) \cos\psi + \frac{i}{2}(R - \frac{1}{R}) \sin\psi, \quad R > r, \ -\pi < \varphi \leq \pi. \tag{27}$$

In analogy to (19) and (20), it follows that

$$d_k := T_{k+1/2}(c) = A_k \cos(k + \frac{1}{2})\psi + iB_k \sin(k + \frac{1}{2})\psi, \tag{28}$$

where

$$A_k := \frac{1}{2}(R^{k+1/2} + \frac{1}{R^{k+1/2}}) \quad \text{and} \quad B_k := \frac{1}{2}(R^{k+1/2} - \frac{1}{R^{k+1/2}}). \tag{29}$$

Based on Rivlin and Shapiro's characterization [10] of the optimal solution of general linear Chebyshev approximation problems, we next derive a simple criterion for the polynomial (26) to be best possible in (15). Note that the extremal points of $v_n(z; c)$ are just the z_l, $l = -n, \ldots, n$, stated in (22). By applying the theory [10] to (15), (26), and by using (23), we obtain the following

Criterion. *$v_n(z;c)$ is the unique optimal polynomial for (15) iff there exist nonnegative real numbers $\sigma_{-n}, \sigma_{-n+1}, \ldots, \sigma_n$ (not all zero) such that*

$$\sum_{l=-n}^{n} \sigma_l(-1)^l w(z_l) q(z_l) = 0 \quad \textit{for all} \quad q \in \Pi_n \quad \textit{with} \quad q(c) = 0. \tag{30}$$

Clearly, it suffices to check (30) for the polynomials

$$q(z) \equiv V_k(z) - V_k(c), \quad k = 1, 2, \ldots, n.$$

With (17) and (28), this leads to the following equivalent formulation of (30):

$$\sum_{l=-n}^{n} \sigma_l(-1)^l (d_0 T_{k+1/2}(z_l) - d_k T_{1/2}(z_l)) = 0, \quad k = 1, 2, \ldots, n. \tag{30'}$$

It turns out that there are simple formulae for all real solutions σ_l of (30′). The trick is to use the ansatz

$$\sigma_l = \sum_{j=0}^{n} (\mu_j \cos(j\varphi_l) + \nu_j \sin(j\varphi_l)), \quad \varphi_l = \frac{2l\pi}{2n+1}, \quad l = -n, \ldots, n, \tag{31}$$

where $\mu_j, \nu_j \in \mathbb{R}$, $j = 0, \ldots, n$. Note that such a representation (31) is possible for any collection of $\sigma_{-n}, \ldots, \sigma_n \in \mathbb{R}$. Now we insert (31) into (30′) and rewrite $T_{k+1/2}(z_l)$ and $T_{1/2}(z_l)$ in the form (19). Then, a routine calculation, making repeatedly use of Lemma 1, shows that (30′) reduces to the equations

$$\mu_k a_{n-k} - i\nu_k b_{n-k} - (d_{n-k}/d_0)(\mu_n a_0 - i\nu_n b_0) = 0, \quad k = 1, 2, \ldots, n-1, \tag{32a}$$

and

$$2\mu_0 a_n - (d_n/d_0)(\mu_n a_0 - i\nu_n b_0) = 0. \tag{32b}$$

By determining all real solutions μ_k, ν_k of the linear system (32a,b) and with (31), one easily verifies that all real numbers satisfying (30′) are given by $\sigma_l = \tau\sigma_l^\star$, $l = -n, \ldots, n$ with $\tau \in \mathbb{R}$ arbitrary and $\sigma_l^\star$ defined in (33). Hence, in view of the Criterion, we have proved the following

Theorem 2. *$v_n(z;c)$ is the unique optimal polynomial in (15) iff the numbers*

$$\sigma_l^\star := \frac{1}{2}\frac{|d_n|^2}{a_n} + \sum_{k=1}^{n}\Big(\frac{\operatorname{Re}(d_{n-k}\overline{d_n})}{a_{n-k}}\cos(k\varphi_l) + \frac{\operatorname{Im}(d_{n-k}\overline{d_n})}{b_{n-k}}\sin(k\varphi_l)\Big),$$
$$l = -n, -n+1, \ldots, n-1, n, \tag{33}$$

are either all nonnegative or all nonpositive. Here a_k, b_k, d_k, and φ_l are defined in (20), (28), and (22).

The numbers (33) are positive whenever R/r is sufficiently large. In particular, in the next section we will prove the following

Theorem 3. *Let $c = c_R(\psi)$ with $R > r > 1$ (cf. (27)). Then:*
a)

$$E_{n+1/2}(r,c) \le \frac{r^{n+1/2} + 1/r^{n+1/2}}{\sqrt{(R^{n+1/2} + 1/R^{n+1/2})^2 - 4\sin^2(n+1/2)\psi}}. \tag{34}$$

b) If $R \ge r(33r-1)/(r-1)$, then $v_n(z;c)$ is the unique optimal polynomial for (15) and equality holds in (34).

For the case that c in (15) is real, we have the following sharper result.

Theorem 4. *Let $r > 1$ and $c \in \mathbf{R}$. If (i) $c \ge r + 1/r - 1/2$ or (ii) $c \le -r$, then $v_n(z;c)$ is the unique optimal polynomial for (15) and*

$$E_{n+1/2}(r,c) = \begin{cases} \dfrac{r^{n+1/2} + 1/r^{n+1/2}}{R^{n+1/2} + 1/R^{n+1/2}} & \text{in case (i)} \\[2ex] \dfrac{r^{n+1/2} + 1/r^{n+1/2}}{R^{n+1/2} - 1/R^{n+1/2}} & \text{in case (ii)} \end{cases}.$$

Remark 2. In contrast to the case $r = 1$, for $r > 1$, the polynomial $v_n(z;c)$ is not best possible in (15) for all $c \in \mathbf{R} \setminus \mathcal{E}_r$. Indeed, numerical tests show that among the corresponding numbers (33), in general, positive and negative σ_l^* occur if c is very close to $\mathcal{E}_r$.

Finally, we note that Theorem 3 is analogous to the following result for the unweighted approximation problem (14).

Theorem A. (Fischer, Freund [3]). *Let $c = c_R(\psi)$ with $R > r > 1$ (cf. (27)). Then:*
a)

$$E_n(r,c) \le \frac{r^n + 1/r^n}{R^n + 1/R^n}. \tag{35}$$

b) If $R \ge r(73r^4 - 1)/(r^4 - 1)$, then

$$p_n(z;c) \equiv \frac{(R^n - 1/R^n)T_n(z) + 2i\sin(n\psi)}{(R^n - 1/R^n)T_n(c) + 2i\sin(n\psi)}$$

is the unique optimal polynomial for (14) and equality holds in (35).

Remark 3. For $n = 1$, (14) was solved completely by Opfer and Schober [7]. From their result, one can deduce (see [3]) that, for the case $n = 1$, the statement in part b) of Theorem A is true for all $R > r \ge 1$.

Clearly, in view of (13), Theorem 1 is an immediate consequence of Theorem 3, Theorem A, and Remark 3. The proofs of Theorem 3 and 4 will be given in the next section.

4. Proofs of Theorem 3 and 4

Proof of Theorem 3. With (17) and (21), it follows that

$$E_{n+1/2} \leq \left\| \frac{w(z)V_n(z)}{w(c)V_n(c)} \right\|_{\mathcal{E}_r} = \frac{a_n}{|T_{n+1/2}(c)|}.$$

By (20) and (28) (both with $k = n$), the right-hand side is just the upper bound in (34). We now turn to part b). Using (28) and (29), it follows that

$$(36) \qquad \begin{aligned} |\operatorname{Re}(d_k\overline{d_n})| &\leq A_kA_n + B_kB_n \leq R^{n+k+1}, \\ |\operatorname{Im}(d_k\overline{d_n})| &\leq A_kB_n + B_kA_n \leq R^{n+k+1}, \end{aligned}$$

and

$$(37) \qquad |d_n|^2 \geq A_n^2 - 1 \geq \frac{1}{4}R^{2n+1}(1 - 2/R^{2n+1}).$$

Let $\sigma_l^\star$ be given by (33). With (36), (37), and (20), we obtain the lower bound

$$(38) \qquad \sigma_l^\star \geq \frac{1}{4}\Big(\frac{R^2}{r}\Big)^n \frac{R\sqrt{r}}{r+1}(1 - \frac{2}{R^{2n+1}}) - 4R^{n+1}\sqrt{r}\sum_{k=0}^{n-1}\Big(\frac{R}{r}\Big)^k \frac{r^{4k+1}}{r^{4k+2}-1}.$$

Now assume that $R \geq r(33r-1)/(r-1)$. With

$$1 - \frac{2}{R^{2n+1}} \geq \frac{1}{2} \quad \text{and} \quad \frac{r^{4k+1}}{r^{4k+2}-1} \leq \frac{r}{r^2-1}, \quad k = 0, 1, \ldots \quad ,$$

we deduce from (38) that

$$\sigma_l^\star \geq \frac{1}{8}\Big(\frac{R^2}{r}\Big)^n \frac{R\sqrt{r}}{(r+1)(R-r)}(R - r - 32\frac{r^2}{r-1}) \geq 0.$$

In view of Theorem 2, this concludes the proof of Theorem 3. ■

Proof of Theorem 4. First we consider the case (i), i.e. assume that

$$(39) \qquad c = \frac{1}{2}(R + \frac{1}{R}) \geq r + \frac{1}{r} - \frac{1}{2}.$$

Then $\psi = 0$ in (28), and the representation (33) reduces to

$$\sigma_l^\star = A_n(\frac{1}{2}\frac{A_n}{a_n} + \sum_{k=1}^{n}\frac{A_{n-k}}{a_{n-k}}\cos(k\varphi_l)), \quad l = -n, \ldots, n.$$

It follows that $\sigma_l^\star = A_n t(\varphi_l)$ where t is the trigonometric polynomial (25) with

$$(40) \qquad \lambda_0 := \frac{A_n}{a_n} \quad \text{and} \quad \lambda_k := \frac{A_{n-k}}{a_{n-k}}, \quad k = 1, \ldots, n.$$

Therefore, Theorem 2 together with Lemma 2 implies that v_n is best possible in (15) provided that the numbers (40) satisfy the assumptions of Lemma 2. Hence, it remains to verify that the estimates

$$(41) \qquad \frac{A_1}{a_1} \geq 2\frac{A_0}{a_0} \quad \text{and} \quad \frac{A_{k+1}}{a_{k+1}} - 2\frac{A_k}{a_k} + \frac{A_{k-1}}{a_{k-1}} \geq 0, \quad k = 1, \ldots, n-1,$$

hold. It is easily seen that the first condition in (41) is equivalent to (39). A more lengthy, but straightforward, computation shows that (39) also guarantees that the remaining inequalities in (41) are satisfied. We omit the details.
For the case (ii), $c \leq -r$, one proceeds similarly. Now $\psi = \pi$ in (28), and from (33) we obtain

$$\sigma_l^* = B_n(\frac{1}{2}\frac{B_n}{a_n} + \sum_{k=1}^{n} \frac{B_{n-k}}{a_{n-k}} \cos k(\varphi_l + \pi)), \quad l = -n, \ldots, n.$$

By applying Lemma 2, this time with

$$(42) \qquad \lambda_0 := \frac{B_n}{a_n} \quad \text{and} \quad \lambda_k := \frac{B_{n-k}}{a_{n-k}}, \quad k = 1, \ldots, n,$$

and Theorem 2, we conclude that v_n is the optimal polynomial for (15) if the assumptions of Lemma 2 are satisfied. A lengthy computation shows that the condition $c \leq -r$ indeed implies that the numbers (42) fulfill the required inequalities. Again, details are omitted here. ∎

Acknowledgement. The author would like to thank Dr. Bernd Fischer for performing some numerical experiments which were very helpful for developing the results of Section 3.

References

[1] S. Bernstein, *Sur une classe de polynomes d'écart minimum*, C. R. Acad. Sci. Paris **190** (1930), 237-240.

[2] C. Frappier, Q.I. Rahman, *On an inequality of S. Bernstein*, Can. J. Math. **34** (1982), 932-944.

[3] B. Fischer, R. Freund, *On the constrained Chebyshev approximation problem on ellipses*, J. Approx. Theory (to appear).

[4] R. Freund, St. Ruscheweyh, *On a class of Chebyshev approximation problems which arise in connection with a conjugate gradient type method*, Numer. Math. **48** (1986), 525-542.

[5] N.K. Govil, V.K. Jain, G. Labelle, *Inequalities for polynomials satisfying $p(z) \equiv z^n p(1/z)$.*, Proc. Amer. Math. Soc. **57** (1976), 238-242.

[6] G. Meinardus, *Approximation of Functions: Theory and Numerical methods*, Springer Verlag Berlin, Heidelberg, New York, 1967.

[7] G. Opfer, G. Schober, *Richardson's iteration for nonsymmetric matriccs*, Linear Algebra Appl. **58** (1984), 343-361.

[8] G. Pólya, G. Szegö, *Aufgaben und Lehrsätze aus der Analysis*, Vol. I., 4th ed., Springer Verlag Berlin, Heidelberg, New York, 1970.

[9] Q.I. Rahman, *Some inequalities for polynomials*, Proc. Amer. Math. Soc. **56** (1976), 225-230.

[10] T.J. Rivlin, H.S. Shapiro, *A unified approach to certain problems of approximation and minimization*¡ J. Soc. Indust. Appl. Math. **9** (1961), 670-699.

[11] W. Rogosinski, G. Szegö, *Über die Abschnitte von Potenzreihen, die in einem Kreise beschränkt bleiben*, Math. Z. **28** (1928), 73-94.

[12] V.I. Smirnov, N.A. Lebedev, *Functions of a Complex Variable*, Iliffe Books, London, 1968.

[13] G. Szegö, *Orthogonal polynomials*, Amer. Math. Soc. Colloq. Publ., Vol. 23. Providence, R. I.: Amer. Math. Soc., 4th ed., 1975.

Received: May 15, 1989

Computational Methods and Function Theory
Proceedings, Valparaíso 1989
St. Ruscheweyh, E.B. Saff, L. C. Salinas, R.S. Varga (*eds.*)
Lecture Notes in Mathematics **1435**, pp. 57–70

Conformal Mapping and Fourier-Jacobi Approximations

David M. Hough[1]
IPS, ETH-Zentrum, CH-8092 Zürich

Abstract. Let f denote the conformal map of a domain interior (exterior) to a closed Jordan curve onto the interior (exterior) of the unit circle. In this paper, we explain how the corner singularities of the of the derivative of the boundary correspondence function can be represented by Jacobi weight functions, and study the convergence properties of an associated Fourier-Jacobi method for approximating this derivative. The practical significance of this work is that some of the best known methods for approximating f are based on integral equations for either the boundary correspondence function or its derivative.

1. Introduction

In this section the mapping problem is defined and a particular boundary integral representation for the conformal map is introduced. The contents of the rest of the paper are outlined at the end of this section.

Let $\partial\Omega := \bigcup_{k=1}^{N} \Gamma_k$ denote a piecewise analytic Jordan curve in the complex plane whose component analytic arcs $\Gamma_1, \Gamma_2, \ldots, \Gamma_N$ are defined by

$$\Gamma_k := \{z : z = \zeta_k(t),\ -1 \le t \le 1\}, \quad k = 1, 2, \ldots, N,$$

where ζ_k is analytic on a domain containing $[-1, 1]$ and satisfies

$$\zeta_k'(t) \neq 0, \quad -1 \le t \le 1, \quad k = 1, 2, \ldots, N.$$

It is assumed that these arcs are numbered consecutively, so that $\zeta_k(1) = \zeta_{k+1}(-1)$ with $\zeta_{N+1} \equiv \zeta_1$, that the positive direction on each Γ_k keeps int $\partial\Omega$ on the left and that $0 \in$ int $\partial\Omega$. Here, int $\partial\Omega$ and ext $\partial\Omega$ denote respectively the bounded and unbounded components of the plane whose common boundary is the curve $\partial\Omega$. Every corner point

[1]On leave of absence from the Department of Mathematics, Coventry Polytechnic, UK

on $\partial\Omega$ must be a member of $\{\zeta_k(1)\}_{k=1}^N$ but it is not necessary that every member of $\{\zeta_k(1)\}_{k=1}^N$ be a corner point since, for example, Γ_k and Γ_{k+1} could be subarcs of a single analytic arc.

Let Ω denote one of the simply-connected domains int $\partial\Omega$ or ext $\partial\Omega$ and let f denote the conformal map $f : \Omega \to D$ where D is either the interior or exterior of the unit circle, i.e.

$$D := \begin{cases} \{w : |w| < 1\} & \text{if } \Omega \equiv \text{int } \partial\Omega, \\ \{w : |w| > 1\} & \text{if } \Omega \equiv \text{ext } \partial\Omega. \end{cases}$$

For the case $\Omega \equiv$ int $\partial\Omega$ it is assumed without loss of generality that

$$f(0) = 0,$$

and, similarly, for the case $\Omega \equiv$ ext $\partial\Omega$ that

$$f(\infty) = \infty\,.$$

In the latter case, in order that f be one-to-one on Ω, f must have a simple pole at ∞. The orientation of the domain D is conventionally fixed by requiring that $f'(z)$ be real and positive either at the point $z = 0$, in the case $\Omega \equiv$ int $\partial\Omega$, or at the point $z = \infty$, in the case $\Omega \equiv$ ext $\partial\Omega$. The inner radius , b, of int $\partial\Omega$ and the capacity, c, of $\partial\Omega$ are then defined by

$$b := [f'(0)]^{-1} > 0 \quad , \Omega \equiv \text{int } \partial\Omega, \tag{1}$$

and

$$c := [f'(\infty)]^{-1} > 0 \quad , \Omega \equiv \text{ext } \partial\Omega.$$

It is well known that the function f exists uniquely, is analytic almost everywhere in $\overline{\Omega}$ and is continuous on $\partial\Omega$, its only singularities at finite points in $\overline{\Omega}$ being branch point singularities at corner points on $\partial\Omega$.

The boundary correspondence function θ_k associated with the arc Γ_k is defined by

$$\theta_k(t) := \arg(f \circ \zeta_k(t)), \tag{2}$$

where arg must be defined so that each θ_k is continuous on $[-1, 1]$. The functions $\{\theta_k\}_{k=1}^N$ completely define the map f and they, or their derivatives, are the fundamental quantitites which are to be approximated in a number of integral equation methods for the numerical determination of f; see Henrici [3, §16.6-7]. If $z = \zeta_k(t) \in \partial\Omega$ then it follows immediately from (2) that

$$f \circ \zeta_k(t) = \exp(i\theta_k(t)), \tag{3}$$

whilst if $z \in \Omega$ then there are various boundary integral formulations that may be used for the calculation of $f(z)$.

The boundary integral formulation considered here is

$$f(z) = \begin{cases} z \exp(i\,\omega - K(z)) & \text{if } \Omega \equiv \text{int } \partial\Omega, \\ c^{-1} \exp(K(z)) & \text{if } \Omega \equiv \text{ext } \partial\Omega, \end{cases} \tag{4}$$

where ω is a real constant, $K : \Omega \to \mathbf{C}$ is defined by

$$K(z) := \sum_{k=1}^{N} \int_{-1}^{1} \nu_k(t) \log(z - \zeta_k(t))\, dt \tag{5}$$

and $\nu_k : [-1, 1] \to \mathbf{R}$ is defined by

$$\nu_k(t) := \frac{\theta_k'(t)}{2\pi} ; \tag{6}$$

see [2], [3, §16.6] and [5] for further discussion of representations similar to (4). The domain of definition of K may be extended to $\overline{\Omega}$ provided the correct branch is taken for the logarithm appearing in (5). That is, if $z \notin \Gamma_k$ then, as a function of the parameter t, $\log(z - \zeta_k(t))$ must be continuous on $[-1, 1]$ whilst if $z = \zeta_k(\tau) \in \Gamma_k$ then

$$\log(\zeta_k(\tau) - \zeta_k(t)) := \begin{cases} \lim\limits_{z \to \zeta_k(\tau), z \in \Omega} \log(z - \zeta_k(t)) & \text{if } \tau \neq t, \\ 0 & \text{if } \tau = t. \end{cases} \tag{7}$$

Hence, the formula (4) may be used to calculate $f(z)$ for all $z \in \overline{\Omega}$. Note that, in view of (1), ω satisfies

$$\log b + i\omega = K(0) \tag{8}$$

and that, since $|f(z)| = 1$ for all $z \in \partial\Omega$, it follows from (4) that

$$c = \exp(\operatorname{Re}\{K(\hat{\zeta})\}) \tag{9}$$

for any chosen point $\hat{\zeta} \in \partial\Omega$. Also note that an immediate consequence of the definition (6) is that the functions $\{\nu_k\}_{k=1}^{N}$ satisfy

$$\sum_{k=1}^{N} \int_{-1}^{1} \nu_k(t)\, dt = 1 . \tag{10}$$

In the next section it is explained how the presence of corners at the ends of the arc Γ_k induces end point singularities in the corresponding density function ν_k. In particular it is shown that these singularities are represented by a Jacobi weight function

$$w_k(t) = (1 - t)^{\alpha_k}(1 + t)^{\beta_k} ,$$

where α_k , β_k are related to the interior angles of Ω at $\zeta_k(\pm 1)$. Also in §2 the Fourier-Jacobi polynomial approximations to ν_k / w_k are introduced and it is shown that these approximations converge almost uniformly on $[-1, 1]$.

In §3 it is proved that the Fourier-Jacobi partial sums produce a uniformly convergent sequence of approximations to the map f. In §4 a brief outline of the collocation method for the approximate solution of Symm-type integral equations is given and it is shown how this may be viewed as a method for estimating the Fourier-Jacobi partial sums.

2. Singularities of $\{\nu_k\}_{k=1}^N$ and Fourier-Jacobi polynomial approximations

In order to avoid the unnecessary use of subscripts in this section, we let Γ, ζ, θ and ν denote respectively a typical arc from the set $\{\Gamma_k\}_{k=1}^N$ and the associated functions ζ_k, θ_k and ν_k.

From (3) and (6), ν can be expressed directly in terms of f as

$$\nu(t) = -\frac{i(f\circ\zeta)'(t)}{2\pi f\circ\zeta(t)}.$$

Since $f\circ\zeta$ is analytic and non-zero on the open interval $(-1,1)$ it follows that ν is also analytic on $(-1,1)$. However, since the arc end points $\zeta(\pm 1)$ are usually corner points of $\partial\Omega$, $f\circ\zeta$ and hence ν are not usually analytic at ± 1. Using the results of Lehman [8], who derives the asymptotic expansion of the map f in the vicinity of a corner formed by analytic arcs, we have previously established the corresponding asymptotic expansion for the density ν in the vicinity of the end points ± 1; see Hough and Papamichael [6].

In order to describe this expansion, let $\lambda\pi$ and $\mu\pi$ denote the angles interior to Ω at the points $\zeta(1)$ and $\zeta(-1)$ respectively and define

$$\alpha := -1 + \lambda^{-1}\,, \quad \beta := -1 + \mu^{-1}.$$

It is assumed always that $\{\lambda, \mu\} \in (0,2)$ so that

$$\{\alpha, \beta\} \in (-\tfrac{1}{2}, \infty)\,.$$

Then, from the results given in [6] it follows that there exists a number δ with $0 < \delta < 1$ such that

$$\nu(t) = \begin{cases} (1-t)^\alpha (a\psi_\lambda(1-t) + \chi^-(1-t)) & \text{if } 1-\delta \le t \le 1, \\ (1+t)^\beta (b\psi_\mu(1+t) + \chi^+(1+t)) & \text{if } -1 \le t \le -1+\delta, \end{cases} \tag{11}$$

where a, b are constants, $\chi^\pm$ are differentiable on $[0,\delta]$ and satisfy $\chi^\pm(0) > 0$ and ψ_γ is defined explicitly on $[0,\delta]$ for any $\gamma \in (0,2)$ by

$$\psi_\gamma(t) := \begin{cases} t^{1/\gamma} & \text{if } \gamma > 1\,, \\ t\log t & \text{if } \gamma = 1\,, \\ 0 & \text{if } \gamma < 1\,, \end{cases}$$

and satisfies $\psi_\gamma(0) = 0$. Clearly, ψ_γ also satisfies a Hölder condition of order $\ell(\gamma)$ on $[0,\delta]$ where

$$\ell(\gamma) := \begin{cases} \gamma^{-1} & \text{if } \gamma > 1\,, \\ 1-\epsilon & \text{for arbitrarily small } \epsilon > 0 \text{ if } \gamma = 1\,, \\ 1 & \text{if } \gamma < 1\,, \end{cases} \tag{12}$$

so that $\frac{1}{2} < \ell(\gamma) \le 1$.

From (11) it is clear that on each arc Γ it is always possible to express ν as the product

$$\nu = w\phi \tag{13}$$

where

$$w(t) := (1-t)^{\alpha}(1+t)^{\beta} \tag{14}$$

is the classical Jacobi weight function and ϕ is a smoother function than ν. The decomposition (13) is useful in that in order to construct approximations to ν, which, from (11), lies in $L^2[-1,1]$, it is only necessary to approximate ϕ which is Hölder continuous, as follows.

Proposition 2.1. *With notation as above, for each arc Γ the quotient function $\phi := \nu/w$ satisfies $\phi \in H^{\alpha^*}[-1,1]$ with $\alpha^* \in (\frac{1}{2}, 1]$ defined by*

$$\alpha^* := \begin{cases} \min(\lambda^{-1}, \mu^{-1}) & \textit{if } \mu > 1 \textit{ or } \lambda > 1\,, \\ 1 & \textit{if } \lambda < 1 \textit{ and } \mu < 1\,, \\ 1-\epsilon & \textit{for arbitrarily small } \epsilon > 0 \textit{ otherwise.} \end{cases}$$

Proof. From (11), (13) and (14) it is clear that if $t \in [1-\delta, 1]$ then

$$\phi(t) = (1+t)^{-\beta}(a\psi_\lambda(1-t) + \chi^-(1-t)).$$

Sums and products of Hölder continuous funtions are also Hölder continuous. Both $(1+t)^{-\beta}$ and χ^- are in $H^1[1-\delta,1]$ and hence, since $\min(\ell(\lambda),1) = \ell(\lambda)$, it follows that $\phi \in H^{\ell(\lambda)}[1-\delta,1]$. Similarly, $\phi \in H^{\ell(\mu)}[-1,-1+\delta]$ and $\phi \in H^1[-1+\delta, 1-\delta]$. Hence $\phi \in H^{\alpha^*}[-1,1]$ where $\alpha^* := \min(\ell(\mu), 1, \ell(\lambda))$. From (12), this definition of α^* is equivalent to that stated in the proposition. ∎

Naturally, if $\lambda < 1$ and $\mu < 1$ then it can be deduced from the expansion in [6] that a certain order of derivative of ϕ is also Hölder continuous on $[-1,1]$, the order and corresponding Hölder index being determined by λ and μ.

The present work is concerned with the possibility of constructing polynomial approximations to ϕ. Proposition 2.1 guarantees that ϕ is smooth enough for there to exist a sequence of polynomial approximations that converges uniformly to ϕ on $[-1,1]$. The presence of the Jacobi weight function in the decomposition (13) naturally suggests that for numerical work it may be convenient to use Jacobi polynomials as basis functions in any polynomial approximation to ϕ, and this certainly turns out to be the case for the numerical method outlined in §4. Moreover, it is also natural to consider the properties of the Fourier-Jacobi polynomial approximation of degree n defined by

$$S_n := \sum_{m=0}^{n} \langle \phi, p_m \rangle p_m, \tag{15}$$

where p_m is the orthonormal Jacobi polynomial of degree m associated with the weight (14) and the inner product is defined by

$$\langle \phi, \psi \rangle := \int_{-1}^{1} w\phi\overline{\psi}\, dt\,. \tag{16}$$

The equiconvergence theorem of Szegö [14, Theorem 9.1.2] may be used to prove that the sequence $\{S_n\}_{n=1}^{\infty}$ converges almost uniformly to ϕ on $[-1,1]$, as follows.

Proposition 2.2. *The Fourier-Jacobi approximation S_n defined by (15) satisfies*

$$\lim_{n\to\infty} |\phi(t) - S_n(t)| = 0$$

for all t in $(-1,1)$ and the convergence is uniform on any interval $[-1+\delta, 1-\delta]$, where δ is any fixed number in $(0,1)$.

Proof. Define indices $\alpha^+ > 0$ and $\beta^+ > 0$ by

$$\alpha^+ := \frac{2\alpha+1}{4}, \quad \beta^+ := \frac{2\beta+1}{4},$$

and define $\Phi : [-1,1] \to \mathbf{R}$ by

$$\Phi(t) := (1-t)^{\alpha^+}(1+t)^{\beta^+}\phi(t).$$

Clearly, from Proposition 2.1, Φ is Hölder continuous on $[-1,1]$. Let s_n be the polynomial of degree n formed by the nth partial sum of the Fourier-Chebyshev approximation to Φ on $[-1,1]$; i.e., $s_n \circ \cos$ is the nth partial sum of the classical Fourier cosine series expansion of $\Phi \circ \cos$ on $[-\pi,\pi]$. For any $t \in (-1,1)$,

$$\begin{aligned} \phi(t) - S_n(t) = \; & (1-t)^{-\alpha^+}(1+t)^{-\beta^+}(\Phi(t) - s_n(t)) \\ & + (1-t)^{-\alpha^+}(1+t)^{-\beta^+} s_n(t) - S_n(t) . \end{aligned}$$

Therefore, if δ is any fixed number in (0,1) it follows that

$$\begin{aligned} \max_{t\in[-1+\delta,1-\delta]} |\phi(t) - S_n(t)| \le \; & \delta^{-\frac{\alpha+\beta+1}{2}} \max_{t\in[-1+\delta,1-\delta]} |\Phi(t) - s_n(t)| \\ & + \max_{t\in[-1+\delta,1-\delta]} |(1-t)^{-\alpha^+}(1+t)^{-\beta^+} s_n(t) - S_n(t)| . \end{aligned} \tag{17}$$

But Φ is Hölder continuous on $[-1,1]$ and hence the Dini- Lipschitz criterion ensures that s_n converges uniformly to Φ on $[-1,1]$; see., e.g., Rivlin [11, §3.4]. Also, ϕ satisfies the conditions of Szegö's equiconvergence theorem mentioned above, which states that the second term on the right of (17) tends uniformly to 0 on $[-1+\delta, 1-\delta]$. Hence, given any $\epsilon > 0$ there exists an integer $n_1 = n_1(\epsilon, \delta)$ such that

$$\max_{t\in[-1+\delta,1-\delta]} |\phi(t) - S_n(t)| \le \epsilon$$

for all $n \ge n_1$. This proves the proposition, since uniform convergence on $[-1+\delta, 1-\delta]$ certainly implies pointwise convergence on (-1,1). ∎

The fact that S_n is the best approximation of degree n to ϕ in the norm associated with the inner product (16) is sufficient to guarantee certain uniform convergence properties for the corresponding approximation to the conformal map f, as is explained in the next section.

3. The uniform convergence of approximations to the map f

In this section we revert to the use of subscripts to identify quantities associated with a particular arc Γ_k. A number of preliminary definitions are required, as follows.

Let w_k denote the Jacobi weight function associated with the arc Γ_k and let ϕ_k be the corresponding quotient function $\phi_k := \nu_k/w_k$; see (13), (14). Also, let $L^2_k[-1,1]$ denote the set of complex valued functions g defined on $[-1,1]$ such that $\sqrt{w_k}|g| \in L^2[-1,1]$. L^2_k is a Hilbert space with inner product

$$\langle g,h\rangle_k := \int_{-1}^{1} w_k g\bar{h}\,dt$$

and corresponding norm

$$\|g\|_k := \sqrt{\int_{-1}^{1} w_k |g|^2\,dt}\,.$$

For functions f which are defined on $\overline{\Omega}$ we also intoduce the uniform norms

$$\|f\|_{\overline{\Omega}} := \sup_{z\in\overline{\Omega}} |f(z)|\,, \qquad \|f\|_{\partial\Omega} := \sup_{z\in\partial\Omega} |f(z)|\,.$$

In the case where f is analytic on Ω with a continuous extension to $\partial\Omega$, the maximum modulus principle implies that the above two norms are identical.

Given any $z \in \overline{\Omega}$ let $l_z : \partial\Omega \to \mathbb{C}$ be defined by

$$l_z(\zeta) := \begin{cases} \log(z-\zeta) & \text{if } z \in \overline{\Omega}\,, \zeta \in \partial\Omega\,, z \neq \zeta\,, \\ 0 & \text{if } z = \zeta \in \partial\Omega\,, \end{cases} \tag{18}$$

where the logarithm branch is the same as that described prior to (7), so that $\log(z-\zeta_k(t)) = l_z \circ \zeta_k(t)$. It can be proved that $l_z \circ \zeta_k \in L^2_k$ for all finite $z \in \overline{\Omega}$.

Now let us suppose that each function ϕ_k is approximated by a real polynomial $P_{k,n}$ of degree n and that a corresponding approximation f_n to f is generated from (4) by

$$f_n(z) := \begin{cases} z\exp(i\,\omega_n - K_n(z)) & \text{if } \Omega \equiv \operatorname{int} \partial\Omega, \\ c_n^{-1}\exp(K_n(z)) & \text{if } \Omega \equiv \operatorname{ext} \partial\Omega, \end{cases} \tag{19}$$

where, using the above inner product notation,

$$K_n(z) := \sum_{k=1}^{N} \langle l_z \circ \zeta_k, P_{k,n}\rangle_k\,. \tag{20}$$

The estimate ω_n and approximate inner radius, b_n, are obtained from (8) as

$$\log b_n + i\omega_n = K_n(0) \tag{21}$$

whilst c_n is determined using (9) as

$$c_n = \exp(\operatorname{Re}\{K_n(\hat{\zeta})\})\,. \tag{22}$$

Observe that in the case $\Omega \equiv \operatorname{ext}\partial\Omega$, $l_z \circ \zeta_k \cong \log z$ as $z \to \infty$ so that

$$K_n(z) \cong (\sum_{k=1}^{N} \langle 1, P_{k,n} \rangle_k) \log z \ .$$

Thus, it follows from (19), that in order for $f_n(z)$ to have a simple pole at ∞ the polynomials $P_{k,n}$ must satisfy

$$\sum_{k=1}^{N} \langle 1, P_{k,n} \rangle_k = 1 \ ; \tag{23}$$

i.e., the approximations $\{w_k P_{k,n}\}_{k=1}^{N}$ to $\{\nu_k\}_{k=1}^{N}$ must satisfy the counterpart of condition (10).

Proposition 3.1. *Let $\Omega \equiv \mathrm{int}\partial\Omega$ and let $\{P_{k,n}\}_{n=1}^{\infty}$ be sequences of polynomial approximations that satisfy*

$$\lim_{n\to\infty} \|\phi_k - P_{k,n}\|_k = 0 \ , \quad k = 1, 2, \ldots, N \ . \tag{24}$$

Then

$$\lim_{n\to\infty} \|f - f_n\|_{\overline{\Omega}} = \lim_{n\to\infty} \|1 - \frac{f_n}{f}\|_{\overline{\Omega}} = 0 \ .$$

Proof. It is clear from the definition (19) that f_n is analytic on Ω with continuous extension to $\partial\Omega$. Hence, this same property is shared by both the error $f - f_n$ and the relative error $(f - f_n)/f$, since it is readily shown that the singularity in the relative error at the origin is removable. Therefore, using the maximum modulus principle together with the fact that $|f(z)| = 1$ for all $z \in \partial\Omega$ gives

$$\|f - f_n\|_{\overline{\Omega}} = \|f - f_n\|_{\partial\Omega} = \|1 - \frac{f_n}{f}\|_{\partial\Omega} = \|1 - \frac{f_n}{f}\|_{\overline{\Omega}} \ ;$$

i.e.

$$\|f - f_n\|_{\overline{\Omega}} = \|1 - \frac{f_n}{f}\|_{\overline{\Omega}} = \|1 - \frac{f_n}{f}\|_{\partial\Omega} \ . \tag{25}$$

Equations (4) and (19) imply that

$$1 - \frac{f_n(z)}{f(z)} = 1 - \exp(i(\omega_n - \omega) - (K_n - K)(z)) \ . \tag{26}$$

Also, from (5), (20) and the use of the Cauchy-Schwarz inequality it follows that for $z \in \partial\Omega$

$$\begin{aligned} |(K_n - K)(z)| &= |\sum_{k=1}^{N} \langle l_z \circ \zeta_k, P_{k,n} - \phi_k \rangle_k| \\ &\leq \sum_{k=1}^{N} \|l_z \circ \zeta_k\|_k \|P_{k,n} - \phi_k\|_k \\ &\leq M E_n \ , \end{aligned} \tag{27}$$

where

$$M := \sup_{z\in\partial\Omega} \max_{k=1,\ldots,N} \|l_z \circ \zeta_k\|_k$$

and

(18) $$E_n := \sum_{k=1}^{N} \|P_{k,n} - \phi_k\|_k \, .$$

Similarly, from (8)and (21),

(29) $$\begin{aligned}|\omega_n - \omega| &= |\mathrm{Im}\,\{(K_n - K)(0)\}| \\ &\le |(K_n - K)(0)| \\ &\le M_0 E_n \, ,\end{aligned}$$

where

$$M_0 := \max_{k=1,\ldots,N} \|l_0 \circ \zeta_k\|_k \, .$$

Now, it is readily established that if $|z| \neq 0$ then

$$|1 - \exp(z)| < |z| \exp(|z|)$$

and $|z|\exp(|z|)$ is a monotonically increasing function of $|z|$. Combining this result with (26),(27) and (29) implies that

(30) $$\|1 - \frac{f_n}{f}\|_{\partial\Omega} \le (M + M_0)E_n \exp((M + M_0)E_n) \, ,$$

with equality only if $E_n = 0$. But from (28) it is clear that if $\|\phi_k - P_{k,n}\|_k \to 0$ for $k = 1, \ldots, N$ as $n \to \infty$ then $E_n \to 0$ as $n \to \infty$ which, using (25) and (30), proves the proposition. ∎

The above proposition cannot hold fully for the case $\Omega \equiv \mathrm{ext}\partial\Omega$ since the error $f - f_n$ has a pole at ∞. However, that part of Proposition 3.1 relating to the convergence of the relative error norm carries over to the exterior domain.

Proposition 3.2. *Let $\Omega \equiv \mathrm{ext}\partial\Omega$ and let $\{P_{k,n}\}_{n=1}^{\infty}$ be sequences of polynomial approximations that satisfy the convergence conditions (24) together with the condition (23). Then*

(31) $$\lim_{n\to\infty} \|1 - \frac{f_n}{f}\|_{\overline{\Omega}} = 0 \, .$$

Proof. Since the polynomials $\{P_{k,n}\}_{k=1}^{N}$ satisfy the conditions (23) for each n it follows from (19) that $f_n(z) \cong c_n^{-1} z$ as $z \to \infty$ so that the relative error $(f - f_n)/f$ is analytic at ∞ with

(32) $$1 - \frac{f_n(\infty)}{f(\infty)} = \frac{c^{-1} - c_n^{-1}}{c^{-1}} \, .$$

Since f is analytic and non-zero on $\Omega\backslash\infty$ and f_n is analytic on $\Omega\backslash\infty$ it is clear that the relative error is analytic on Ω and has a continuous extension to $\partial\Omega$. Hence the maximum modulus principle applies to give

$$\|1-\frac{f_n}{f}\|_{\overline{\Omega}}=\|1-\frac{f_n}{f}\|_{\partial\Omega}\,.$$

From this point onwards the details are practically identical to those given in the proof of Proposition 3.1 and need only be outlined very briefly. Corresponding to (26), using (9) and (22), there is

$$1-\frac{f_n(z)}{f(z)}=1-\exp(\mathrm{Re}\{(K-K_n)(\hat{\zeta})\}+(K_n-K)(z))$$

and from this it may be deduced that

$$\|1-\frac{f_n}{f}\|_{\partial\Omega}\leq 2ME_n\exp(2ME_n)\,. \tag{33}$$

■

A number of remarks can be made about the above results, as follows. In the first place, it has been observed in various numerical methods for estimating f for the case $\Omega\equiv\mathrm{ext}\partial\Omega$ that the error in the estimate for c is always of significantly smaller magnitude than the maximum relative error in the estimate for f itself; see [13], [9] and [6]. The simple result (32) goes some way towards explaining this observation since it states that the relative error in c_n^{-1} as an estimate for c^{-1} is in fact the value of an analytic function at an interior point of its domain of analyticity; hence $|(c^{-1}-c_n^{-1})/c^{-1}|<\|(f-f_n)/f\|_{\overline{\Omega}}$. This result is independent of the precise method of calculating f_n, provided only that f_n is analytic on $\Omega\backslash\infty$ and $f_n(z)\cong c_n^{-1}z$ as $z\to\infty$.

Secondly, again for the case $\Omega\equiv\mathrm{ext}\partial\Omega$, it will be clear that if, say, Ω^* is any bounded domain in ext $\partial\Omega$ then it is possible to prove that f_n converges uniformly to f on $\overline{\Omega^*}$, provided the conditions of Proposition 3.2 are satisfied.

Thirdly, in the above Propositions the polynomial approximation to ϕ_k may be allowed to take a different degree, say n_k, for each $k=1,\ldots,N$. Provided (24) and the counterpart of (23) hold, then the corresponding approximation to f will have the uniform convergence properties stated in Propositions 3.1, 3.2 as $n:=\min\{n_1,\ldots,n_N\}\to\infty$.

Finally, the Fourier-Jacobi polynomial $S_{k,n}$ of degree n approximating ϕ_k is

$$S_{k,n}:=\sum_{m=0}^{n}\langle\phi_k,p_{k,m}\rangle_k p_{k,m}\,, \tag{34}$$

where $p_{k,m}$ is the orthonormal Jacobi polynomial of degree m associated with the weight w_k, $k=1,\ldots,N$. It is well known that $S_{k,n}$ is the best polynomial approximation of degree n to ϕ_k in the norm $\|\cdot\|_k$. Hence, the choice $P_{k,n}=S_{k,n}$ for $k=1,\ldots,N$ minimises the quantity E_n defined by (28) and therefore also minimises the bounds appearing in (30) and (33). Moreover, since $\phi_k\in\mathrm{L}_k^2$ and $\{p_{k,n}\}_{n=0}^{\infty}$ is a basis for L_k^2, this choice also satisfies the condition (24).

In theory then, the use of the Fourier-Jacobi polynomial approximations to the quotient functions ϕ_k forms a very satisfactory approach to determining approximations to the map f. In practise, it may be a non-trivial task to devise a numerical scheme for estimating the Jacobi coefficients in (34) and which has proven convergence properties as the scheme is refined.

4. The Symm-Jacobi collocation method

A possible method for estimating the Fourier-Jacobi approximation (34) is via the numerical solution of Symm's integral equations. These equations arise in a natural way from the representation (4) by imposing the condition that $|f(z)| = 1$ for all $z \in \partial\Omega$; i.e.,

$$\text{Re}\{K(z)\} = \left\{ \begin{array}{ll} \log|z| & \text{if } \Omega \equiv \text{int } \partial\Omega, \\ \log c & \text{if } \Omega \equiv \text{ext } \partial\Omega, \end{array} \right\}, z \in \partial\Omega . \tag{35}$$

Let R_z denote the real part of the logarithm function l_z defined by (18); i.e.,

$$R_z(\zeta) := \log|z - \zeta| .$$

Then, with notation as in §3, we may write the equation (35) together with the condition (10) as

$$\begin{aligned} \sum_{k=1}^{N} \langle R_z \circ \zeta_k, \phi_k \rangle_k + \gamma &= b_z \quad , z \in \partial\Omega , \\ \sum_{k=1}^{N} \langle 1, \phi_k \rangle_k &= 1 , \end{aligned} \tag{36}$$

where

$$b_z := \left\{ \begin{array}{ll} \log|z| & \text{if } \Omega \equiv \text{int } \partial\Omega, \\ 0 & \text{if } \Omega \equiv \text{ext } \partial\Omega. \end{array} \right.$$

The integral equation and side condition (36) are to be solved for the quotient functions $\{\phi_k\}_{k=1}^N$ and the constant γ. For a piecewise analytic Jordan curve, the analysis of Gaier [2] shows that the unique Hölder continuous solutions of (36) are given by $\phi_k := \theta_k'/(2\pi w_k)$, $k = 1, \ldots, N$, and the unique value of the scalar γ is defined by

$$\gamma := \left\{ \begin{array}{ll} 0 & \text{if } \Omega \equiv \text{int } \partial\Omega, \\ -\log c & \text{if } \Omega \equiv \text{ext } \partial\Omega; \end{array} \right.$$

see also Reichel [10]. In numerical conformal mapping, integral equations of the first kind with a logarithmic kernel are generally known as Symm-type equations, because this kind of formulation was first used by Symm [12], [13]. In fact the equations (36) are not the same as those originally considered by Symm. The inclusion of the scalar γ and the side condition for the interior mapping problem is a device introduced in a more general context by Hsiao and MacCamy [7] in order to produce a system which has a unique solution irrespective of the scale of the domain Ω. We have seen in Proposition 3.2 that it is essential in the case $\Omega \equiv \text{ext}\, \partial\Omega$ to impose the side condition on a numerical solution in order to produce the correct mapping behaviour at ∞. The equation (36), without the side condition, for the exterior mapping problem is due to Gaier [2] and has the advantage that its solutions $\{\phi_k\}$ are generally in smoother Hölder continuity classes than the corresponding solutions to Symm's original formulation [13]; see [6] for some discussion of this.

The truncation error in the Fourier-Jacobi approximation of degree n to the kernel function $R_z \circ \zeta_k$ of (36) is

$$E^n_{z,k} := R_z \circ \zeta_k - \sum_{m=0}^{n} \langle R_z \circ \zeta_k, p_{k,m} \rangle_k p_{k,m} \,. \tag{37}$$

Writing $\phi_k = S_{k,n} + (\phi_k - S_{k,n})$ in (36) and using (37) together with the fact that $\phi_k - S_{k,n}$ is orthogonal with respect to $\langle \cdot, \cdot \rangle_k$ to all polynomials of degree at most n it follows that

$$\begin{aligned} \sum_{k=1}^{N} \langle R_z \circ \zeta_k, S_{k,n} \rangle_k + \gamma + e^n_z &= b_z \quad, \; z \in \partial\Omega \,, \\ \sum_{k=1}^{N} \langle 1, S_{k,n} \rangle_k &= 1 \,. \end{aligned} \tag{38}$$

Here, the satisfaction of the side condition follows from (34) by using the orthonormality of the Jacobi basis polynomials and e^n_z is defined by

$$e^n_z := \sum_{k=1}^{N} \langle E^n_{z,k}, \phi_k - S_{k,n} \rangle_k \,.$$

Applying the Cauchy-Schwarz inequality to the previous definition gives

$$|e^n_z| \le \sum_{k=1}^{N} \| E^n_{z,k} \|_k \| \phi_k - S_{k,n} \|_k \,.$$

But $\|E^n_{z,k}\|_k \to 0$ and $\|\phi_k - S_{k,n}\|_k \to 0$ for each k as $n \to \infty$ so that $|e^n_z| \to 0$ as $n \to \infty$. This suggests that a simple approach towards estimating the polynomials $\{S_{k,n}\}_{k=1}^N$ is therefore to reduce (38) to a linear algebraic system by collocating at an appropriate number of points on $\partial\Omega$ and ignoring the contribution of e^n_z at these points, as follows.

Let $\{z_i\}_{i=1}^{Nn+N}$ be a set of distinct points on $\partial\Omega$, with $n+1$ points located on each arc Γ_k, $k = 1, \ldots, N$. Then the polynomials of degree n, say $P_{k,n}$, which estimate the Fourier-Jacobi partial sums $S_{k,n}$, $k = 1, \ldots, N$, satisfy the equations

$$\begin{aligned} \sum_{k=1}^{N} \langle R_{z_i} \circ \zeta_k, P_{k,n} \rangle_k + \gamma_n &= b_{z_i} \quad, \; z_i \in \partial\Omega \,, \\ \sum_{k=1}^{N} \langle 1, P_{k,n} \rangle_k &= 1 \,, \end{aligned} \tag{39}$$

where γ_n is an estimate for γ. Using the orthonormal Jacobi polynomials as a basis for the representation of $P_{k,n}$, i.e.,

$$P_{k,n} = \sum_{m=0}^{n} a_{k,m} p_{k,m} , \tag{40}$$

in (39) produces a $(Nn+N+1)\times(Nn+N+1)$ linear algebraic system for the unknowns $a_{k,m}$, $k = 1,\dots,N$, $m = 0,\dots,n$, and γ_n.

It is appropriate to mention here that previous methods which aim to treat the end-point singularities of ν_k via the numerical solution of Symm-type equations have been based on the use of piecewise approximations combined with a variety of techniques for dealing with the singularities. For example, in [6] we construct approximations to ν_k using cubic B-spline basis functions augmented by special singular basis functions with support restricted to subintervals adjacent to the end points on $[-1,1]$. Hoidn [4] considers a technique for the systematic non- analytic reparametrisation of Γ_k in such a way that the modified density is sufficiently smooth to allow the possibility of constructing global spline approximations. Costabel and Stephan [1] analyse the use of linear spline approximations with suitably graded meshes near the ends of $[-1,1]$. It must be said that the latter method is currently the most mathematically respectable since a complete convergence analysis is given. Nevertheless, the orthogonal polynomial approach provides a simple and reasonably elegant means of accomodating end point singularities and it is also relatively straightforward to implement. The preliminary numerical implementation described in [5] produced excellent approximations to f on several test problems and only involved the solution of relatively small linear algebraic systems.

Although there is no convergence analysis currently available for the Symm-Jacobi collocation method, it can still be used with confidence for practical mapping computations since reliable a posteriori error indicators are available to assess the accuracy of f_n. In particular the the error in modulus $|1-|f_n(z)||$ can be computed for $z \in \partial\Omega$ and $f_n(z)$ for $z \in \partial\Omega$ can be compared with the alternative representation obtained via (3).

The most important computational difficulty for the practical implementation of the scheme (39), (40) is that of producing accurate and efficient methods for the estimation of the inner products $\langle R_z \circ \zeta_k, p_{k,m} \rangle_k$ which appear in (37) and which form the matrix elements in the linear algebraic system. Space does not allow us to go into details here, other than to note that the underlying connection between the Jacobi polynomials, the logarithmic kernel and the Jacobi functions of the second kind allows effective use to be made of three term recurrence relations. A full computer implementation of the method, using more accurate quadrature techniques than those used in [5], is currently nearing completion and it is intended that a report describing all computational aspects will be prepared shortly.

References

[1] M. Costabel, E.P. Stephan, *On the convergence of collocation methods for boundary integral equations on polygons*, Math. Comp., **49** (1987), 461-478.

[2] D. Gaier, *Integralgleichungen erster Art und konforme Abbildung*, Math. Z., **147** (1976), 113-129.

[3] P. Henrici, *Applied and Computational Complex Analysis* **III**, Wiley, New York, 1986.

[4] H.P. Hoidn, *A reparametrisation method to determine conformal maps.*, In L.N. Trefethen, editor, Numerical Conformal Mapping, 155-161, North-Holland, Amsterdam, 1986.

[5] D.M. Hough, *Jacobi polynomial solutions of first kind integral equations for numerical conformal mapping*, J. Comput. Appl. Math., **13** (1985), 359-369.

[6] D.M. Hough, N. Papamichael, *An integral equation method for the numerical conformal mapping of interior, exterior and doubly-connected domains*, Numer. Math., **41** (1983), 287-307.

[7] G. Hsiao, R.C. MacCamy, *Solution of boundary value problems by integral equations of the first kind*, SIAM Rev., **15** (1973), 687-705.

[8] R.S. Lehman, *Development of the mapping function at an analytic corner*, Pacific J. Math, **7** (1957), 1437-1449.

[9] N. Papamichael, C.A. Kokkinos, *Numerical conformal mapping of exterior domains*, Comput. Meths. Appl. Mech. Engrg., **31** (1982), 189-203.

[10] L. Reichel, *On polynomial approximation in the complex plane with application to conformal mapping*, Math. Comp, **44** (1985), 425-433.

[11] T.J. Rivlin, *The Chebyshev Polynomials*, Wiley, New York, 1974.

[12] G.T. Symm, *An integral equation method in conformal mapping*, Numer. Math., **9** (1966), 250-258.

[13] G.T. Symm, *Numerical conformal mapping of exterior domains*, Numer. Math., **10** (1967), 437-445.

[14] G. Szegö, *Orthogonal Polynomials*, American Mathematical Society, New York, 1975.

Received: April 6, 1989.

Computational Methods and Function Theory
Proceedings, Valparaíso 1989
St. Ruscheweyh, E.B. Saff, L. C. Salinas, R.S. Varga (*eds.*)
Lecture Notes in Mathematics **1435**, pp. 71–79

Numerical Solutions of the Schiffer Equation[1]

J.A. Hummel

Department of Mathematics, University of Maryland
College Park, Maryland, USA

1. Introduction

Variational methods are powerful tools for studying the properties of classes of univalent functions. These methods can be applied to many specific classes and lead to a *Schiffer Differential Equation* which the extremal function must satisfy. When the functions considered are univalent in the unit disk Δ, the differential equation is usually of the form

$$R(z)\frac{dz^2}{z^2} = Q(w)dw^2 \tag{1}$$

where R and Q are rational functions. This differential equation is an equality between quadratic differentials and its solution often requires the use of properties of these quadratic differentials.

Equation (1) is actually a *functional differential equation.* That is, the equation itself involves parameters which depend on the unknown function which solves the extremal problem. In the past, the equation was often solved by exhibiting all possible solutions of (1) and choosing from these one which actually produced the extreme value. This required that (1) have solutions which could be written in terms of elementary functions which was true if (1) was simple enough, or was a perfect square [2]. More recently, the problems being studied became more complex and integration of either side (or more properly, the square root of either side) of (1) would lead to elliptic or hyperelliptic integrals. With the availability of computers, we may take the attitude that a function defined by such an integral is just as good as one of the *elementary* functions. We merely have to compute it as needed rather than look it up in a table. We of course lose some of the relations between the elementary functions, so it becomes more difficult to select the proper values of the parameters to produce the extremal function.

[1]Work on this paper was supported in part by a grant from the National Science Foundation.

2. An example

The problems can be illustrated by considering a specific example [4] which is just complex enough to show some of the difficulties. Let two non-zero points, a and b, be given in the unit disk Δ. We wish to find a domain $D \subset \Delta$ with $a, b \in D$ and $0 \notin D$ such that the hyperbolic distance between a and b is minimum (or equivalently find such a D for which the Green's function $g(a, b)$ is a minimum). We can also invert the problem and search for a function $f(z)$ analytic, univalent, and non-zero in the unit disk which maps Δ into Δ, and such that $f(0) = a$, $f(r) = b$ where $0 < r < 1$ and r is a minimum. Using the variational method derived in [3], this leads to the Schiffer differential equation of the form (1) for the extremal function $w = f(z)$ with

$$\begin{aligned} R(z) &= \frac{rz(1-r^2)}{(1-rz)(r-z)}, \\ Q(w) &= \frac{1}{w^2}\left[\frac{b^2 b_1}{b-w} - \frac{\overline{bb_1}}{1-\bar{b}w} + \frac{(1-r^2)a}{\overline{a_1}(1-aw)} - \frac{(1-r^2)a^2}{a_1(a-w)}\right] \end{aligned} \tag{2}$$

where $b_1 = 1/f'(r)$, $a_1 = f'(0)$, and we assume without loss of generality that $a > 0$. The unit circle in the z plane is a trajectory of the quadratic differential on the left hand side of (1) since $R(z) < 0$ on this circle. In general, when we consider a class of functions which are univalent in the unit disk, the unit circle in the z plane will be a trajectory. Thus it is convenient to write the left hand side of (1) with the expression dz^2/z^2 separated out.

It further follows from the variational method for this particular example, that $w^2 Q(w)$ is real and negative when $|w| = 1$ and that it must have a zero of order 2 at some point w_0 on the unit circle. It therefore follows that $Q(w)$ must be of the form

$$Q(w) = \frac{c^2 \overline{w_0}(w-w_0)^2}{w(b-w)(a-w)(\bar{b}w-1)(aw-1)}, \tag{3}$$

where $|w_0| = 1$ and c is real. Thus equation (1) involves the known parameters a (real) and b (complex), and the three unknown parameters r (real), c (real), and w_0 ($|w_0| = 1$), reduced from the five unknown real parameters in r, a_1, and b_1, in (2). The reduction results from the properties of $Q(w)$ following from the variational method. This type of reduction is typical of the situation arrived at when the variational method is used.

3. Defining the conformal map

Equation (1) is separable. Each side can be integrated to obtain a multiple valued function. The desired mapping is implicitely determined by the equality between the two integrals. But, how can the unknown parameters be determined? Let us write $d\Omega^2 = R(z)\frac{dz^2}{z^2}$ and then define $\Omega(z) = \int d\Omega$. Apart from the zeros and poles of $d\Omega^2$, $\Omega(z)$ is a locally univalent function which maps trajectory arcs of $d\Omega^2$ to horizontal lines.

For this particular example, we observe from (2) that the line segment from 0 to r is a trajectory of $d\Omega^2$. Thus, the interior of the unit circle less this line segment is a ***Ring Domain*** and if we cut this domain along the conjugate trajectory from r to 1, then $\Omega(z)$ will be univalent in $\Delta_1 = \Delta\backslash[0,1)$ and will map this domain onto the interior of a rectangle R_z in the Ω plane.

In the same way, we can set $d\Omega_w^2 = Q(w)dw^2$ and obtain a multivalued mapping from the w plane into the Ω_w plane. However, the function $w = f(z)$ which solves the desired problem must satisfy the differential equation and hence must map Δ_1 onto a domain in the w plane bounded by trajectory arcs and conjugate trajectory arcs of $d\Omega_w^2$ which must map onto a rectangle R_w in the Ω_w plane, congruent to the rectangle R_z. To solve the problem, we first attempt to find parameters so that the trajectory structures of $d\Omega^2$ and $d\Omega_w^2$ are of the same type. Usually, the left hand side of (1) will have only a small number of possible trajectory structures as in this case where the only possible structure of $d\Omega^2$ is a ring domain. We look for w_0 so that $d\Omega_w^2$ will have this structure also. (Note that since c is real and positive, changes in c will not change the trajectory structure of $d\Omega_w^2$.) When the structure of $d\Omega^2$ is more complicated, the problem of determining the parameters so that $d\Omega_w^2$ has the same structure can be quite difficult, but it usually reduces to finding parameter values so that certain integrals are real.

Further, depending on the class being considered, other conditions will have to be satisfied. Since $f(z) \neq 0$ in Δ in our example, one of the trajectories of $d\Omega_w^2$ must join w_0 and 0. The other two simple poles of $d\Omega_w^2$, at a and b, must be joined by a trajectory γ which is the image under f of the line segment $(0,r)$ in the z plane. Thus, all we need to do is find a value for w_0 so that $\int_\gamma d\Omega$ is real. Then the trajectory from 0 must go to w_0 and the structure of the resulting $d\Omega_w^2$ must be that of a ring domain. Since $|w_0| = 1$, the value of this integral depends on only a single real parameter, and thus this is an easy problem to solve numerically provided we know γ (or a curve homotopic to γ in Δ less the three singularities of $d\Omega_w$ at a, b, and 0). This is particularly easy in this case since $d\Omega_w^2$ contains the factor $\overline{w_0}(w-w_0)^2$ and hence

$$\int_\gamma d\Omega = w_0^{-1/2}\int_\gamma w\,d\Omega_1 - w_0^{1/2}\int_\gamma d\Omega_1 \tag{4}$$

where $d\Omega_1^2 = c^2dw^2/[w(b-w)(a-w)(aw-1)(\bar{b}w-1)]$. Thus, w_0 is determined by the values of the two integrals which do not depend on w_o. The unknown parameter c is real and hence cancels out in the determination of w_0. In other, more general, cases it would be necessary to solve a system of equations in order to make the structures the same. If $d\Omega^2$ is complicated, this might have to be done for a number of possible structures.

The second part of the problem is to find values of the parameters so that the images of Δ in the z plane and $f(\Delta)$ in the w plane are identical Riemann surfaces in the Ω and Ω_w planes respectively. In our example, this only requires that the moduli of the ring domains be the same because the parameter c can then be adjusted so that the sizes of the rectangles are identical. In the z plane, the width of this rectangle is the integral of $d\Omega$ around $|z| = 1$, or twice the integral of $d\Omega$ along the real axis from 0 to r. The height of the rectangle is the imaginary part of the integral of $d\Omega$ from r to 1. The modulus μ_r is the ratio of these and is a monotone increasing function of r. In the w plane, the width of the rectangle is twice the integral of $d\Omega_w$ along γ. The height is the imaginary part of the integral of $d\Omega_w$ along a curve γ' from 0 to a which does not intersect γ. Once w_o is determined, these integrals can be calculated and we can solve for r so that μ_r equals the modulus of the ring domain in the w plane. Thus, we can easily solve the superficially bigger problem of finding the minimum for each given homotopy class of γ.

4. Solving the extremal problem

The final problem is to select the particular solution which solves the extremal problem. In this case, that means to select the proper homotopy class for γ. It can be shown from the theory of quadratic differentials, [5], [6], that this is the class containing the line segment joining a to b (with appropriate changes if a, b, and 0 are collinear). However, even if such information is not available it is often possible to select the proper solution using the available numerical information. In [4] it was shown that only a finite number of homotopy classes are feasible for a given a and b. Thus it is easy to pick out the actual extremal value from these. This type of reasoning would usually be applicable in more complex cases where no theoretical results are available.

In order to solve for the w_0 which makes the integral (4) purely real for a given homotopy class of γ, we must reduce the integrals to manageable form. We observe that any γ joining a and b is homologically equivalent in $\Delta\backslash\{0,a,b\}$ to $n_1\gamma_1 + n_2\gamma_2$ where γ_1 and γ_2 are the line segments joining 0 to a and 0 to b respectively (with appropriate changes if $b > 0$) and n_1 and n_2 are odd integers with signs chosen properly to give the same integral. That is, if we compute

$$I_j = \int_{\gamma_j} w\, d\Omega_1, \quad J_j = \int_{\gamma_j} d\Omega_1, \quad j = 1,2, \tag{5}$$

fixing a sign for each one, then, given γ, there exist a pair of odd integers n_1 and n_2 such that, setting $I = n_1 I_1 + n_2 I_2$ and $J = n_1 J_1 + n_2 J_2$, $\int_\gamma d\Omega = w_0^{-1/2} I - w_0^{1/2} J$. We will call this pair of integers n_1, n_2, the *signature* of the arc γ.

-2.3561945
1, 0, 0
1, 1

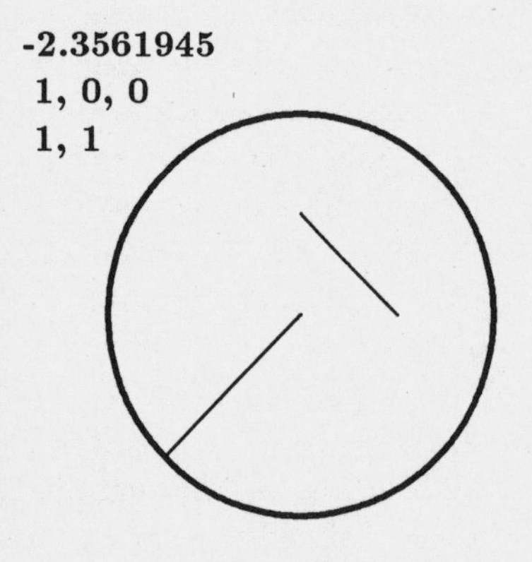

Figure 1

0.7853982
1, 0, 2
1,-1

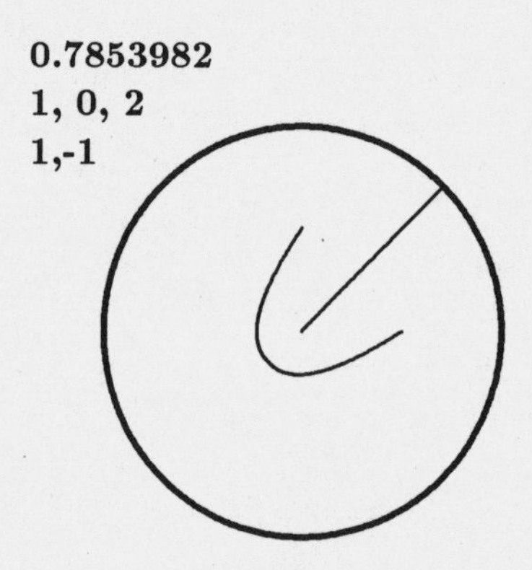

Figure 2

The equation $\mathrm{Im}\int_\gamma d\Omega = 0$ is homogeneous in n_1 and n_2. That is, the pairs n_1, n_2 and kn_1, kn_2 will result in the same w_0 satisfying the above equation. Thus, every relatively prime n_1 and n_2 will define a w_0. Three interesting questions present themselves: First, does every pair of relatively prime odd integers correspond to an actual homotopy class? Second, can there be more than one homotopy class with the same signature n_1, n_2? And third, can a pair of odd integers which are not relatively prime be the signature of some γ? These questions were not considered in [4] since we were interested only in the extremal case, which was known to exist.

First, let us look at some examples. Figures 1 and 2 show the actual trajectory of (3) which joins $|w| = 1$ to 0 for the value of $\theta_0 = \arg w_0$ shown next to it for the case when $a = 0.5$ and $b = 0.5i$. The corresponding trajectories γ joining a to b are also shown. The w_0 were actually computed by solving (4) for the signatures 1,1 and 1,-1 respectively (these signatures are given on the bottom line next to the figure) even though the fact that $|a| = |b|$ means $\arg w_0 = -3\pi/4$ and $\pi/4$ respectively and the trajectory joining 0 to w_0 is a straight line segment.

Since the trajectory of main interest is the one which joins w_0 to 0 (being part of the boundary of $f(\Delta)$) it might seem better to introduce the line segment γ_0 from -1 to 0 along the negative real axis, add $j = 0$ to the definitions in (5) and set $I = I_0+n_1I_1+n_2I_2$ and $J = J_0 + n_1J_1 + n_2J_2$. Any trajectory γ joining $|w| = 1$ to 0 must then define a pair of even integers n_1 and n_2 such that $\mathrm{Im}\int_\gamma d\Omega = \mathrm{Im}[w_0^{-1/2}I - w_0^{1/2}J]$. One can then set this equal to zero and solve for w_0. This process works, but gives much less precise correspondence between the "signature", $1, n_1, n_2$ of this curve and the trajectory structure. (These *three integer* signatures are also shown in the figures.) Note that if we *slide* the end point of the arc joining $|w| = 1$ to 0 in Fig. 1 clockwise from θ_0 to π, the signature would become 1,2,2. In fact, upon calculation, the signatures 1,2,2 and 1,0,0 both result in the same θ_0. (Observe that this process of sliding the initial point of γ_0 along the unit circle does not change the signature of the curve joining a to b.) Fig. 2 is for signatures 1,-1 or 1,0,2. The signature 1,2,0 obtained by decreasing θ_0 to $-\pi$ also gives the same θ_0.

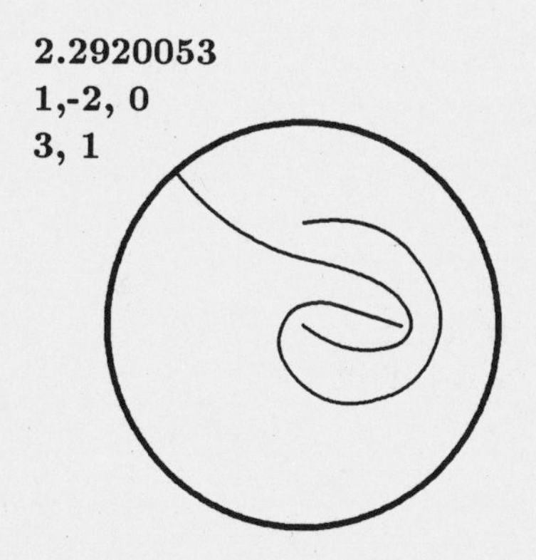

Figure 3

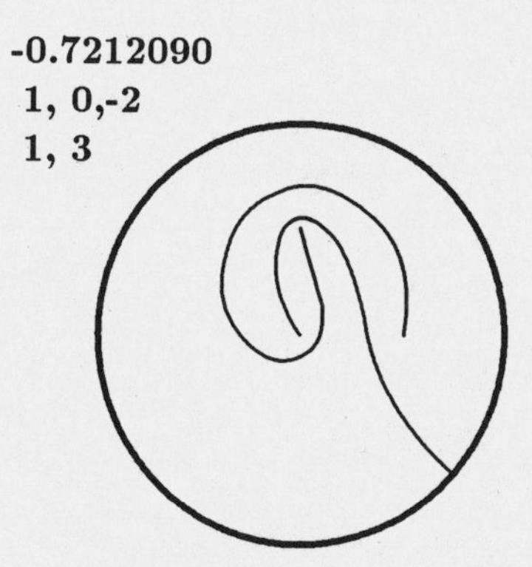

Figure 4

Wrap the γ in Figure 2 once more around the circle, decreasing θ_0 to -3π, resulting in a *three integer* signature 1,4,2. Calculation of θ_0 for this signature gives $\theta_0 = 2.2920053$ with the results shown in Figure 3. The trajectories defined by the differential equation are in a different homotopy class, with signatures 1,-2,0 and 3,1. Starting with signature 1,2,0 and increasing θ_0 to 3π would give a signature 1,2,4. This defines $\theta_0 = -0.7212909$ with the results shown in the fourth figure. This too is actually in a new homotopy class, with signature 1,3, and 1,0,-2 (or 1,2,4).

Thus, use of the signature of the curves joining $|w| = 1$ to 0 seems to cause difficulty. However, the curves joining a to b are much better behaved.

Theorem 1. *The set of possible homotopy classes of simple arcs γ joining a and b in $\Delta\backslash\{0,a,b\}$ is in one to one correspondence with the set of relatively prime odd integers n_1, n_2 with $n_1 > 0$.*

Proof. Any γ can be viewed as starting at the point a and proceeding first to either 0 or b. If to b, we can deform it from a to 0 and then to b. In either case, we assign a positive sign to the first segment of the path. If it continues to b, we assign a plus 1 to the path from 0 to b when the path does not go around 0. Thus the line segment from a to b has signature 1,1 as in Figure 1.

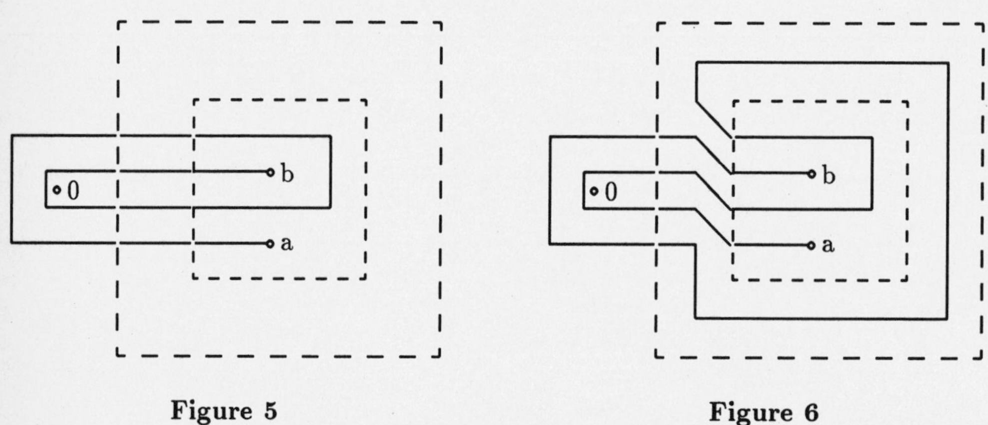

Figure 5 **Figure 6**

Every simple arc joining a and b in $\Delta\backslash\{0,a,b\}$ has a signature n_1, n_2 with n_1 and n_2 odd and $n_1 > 0$. We need only show that each relatively prime odd pair n_1, n_2 with $n_1 > 0$ determines an arc and that there are no arcs for which the signature is not a relatively prime pair. The Dehn-Thurston theorem [1] asserts that the arc γ can be represented up to homotopy by a special system of arcs in a decomposition of the region into two *pairs of pants* and a *belt* as shown symbolically in Fig. 5 where we decompose $\Delta\backslash\{0,a,b\}$ into three regions by two disjoint Jordan curves containing a and b in their interior and 0 in their exterior. These curves are represented by the dashed lines in Fig. 5. (The second *leg* of the outer *pair of pants* is the boundary of the unit disk, and is not shown.) Except for the special case of the homotopy class of the line segment joining a to b (which will be just that), the Dehn-Thurston theorem tells us that the structure in the inner region consists of two arcs joining a and b to the *belt* and some number, say $k-1$, of disjoint arcs with both end points on the *belt*, separating a from b. The outer region will contain k disjoint arcs with both end points on the *belt*, separating 0 from $|w| = 1$. The two sides will be connected by $2k$ disjoint arcs crossing the belt. Figure 5 shows the case of $k = 2$ with direct connections across the belt, resulting in a signature of 1,-3. For a general k, direct connection across the belt defines a curve with signature $1, -(2k-1)$.

The only remaining variable is the number of *Dehn Twists* in the belt. This can be any integer, positive or negative. Fig. 6 shows the result of a Dehn Twist of +1. Each arc crossing the belt has been moved up one position on the left. The arc connecting to the *top* position on the left goes around the belt to connect to the bottom left position. The signature is now 3,-1. That is, each integer in the signature has been increased by two. Another Dehn Twist would similarly result in a signature of 5,1. It is easy to see that the result of each Dehn Twist is equivalent to adding an arc going completely around the points a and b. Half of this is the new arc around the belt, while the other half is made up of the totality of the rest of the shifts. A twist in the reverse direction would subtract two from each number in the signature. If n_1 becomes negative, we would no longer have the standard form for n_1, n_2, but the results remain valid. (For some initial configurations, the results of the positive or negative twists will be reversed, but in every case, a twist in one direction will add two while the reverse twist will substract two.)

Let n_1, n_2 be any pair of odd integers with $n_1 > 0$. By interchanging the roles of a and b if necessary, we may assume without loss of generality that $n_2 \le n_1$. Set $j = (n_1 - 1)/2$ and substract $2j$ from both n_1 and n_2 to give the pair $1, m$ where $m = n_2 - n_1 + 1$. Then $m \le 1$. Suppose first that $m < 0$. Letting $k = (1-m)/2$ and making direct connections across the belt as described above there exists a curve joining a to b with signature $1, m$. We now make a Dehn twist of $+j$ to obtain a configuration with signature n_1, n_2. The question is, does this configuration represent a single arc joining a to b?

We represent the junctures of the arcs on the right and left sides of the belt by R_i and L_i respectively, where i is an integer $0 \le i \le 2k-1$ and the arcs are numbered from the bottom to the top. Thus R_0 and R_k are the end points of the arcs from the points a and b respectively. In the left hand region an arc connects L_i to L_{2k-1-i} for each i, and on the right R_i is connected to R_{2k-i} except when $i = 0$ or k. Now a Dehn shift of $+j$ results in an arc from R_i to R_{2j+1+i} (mod $2k$) by

$$R_i \to L_{i+j} \to L_{2k-1-i-j} \to R_{2k-1-i-2j} \to R_{2j+1+i}.$$

But $2j + 1 = n_1$, so after the twist of j, we have the arc defined by

$$a \to R_0 \to R_{n_1} \to R_{2n_1} \to R_{3n_1} \to \ldots \to R_{pn_1} \to b$$

which terminates when $pn_1 \equiv k \pmod{2k}$. This certainly holds when $p = k$, since n_1 is odd, but may hold for a smaller p. Indeed if $(n_1, k) = q$, then let $p = k/q$. We see that $pn_1 \equiv k \pmod{2k}$ and this will be true for no smaller p. If n_1 and n_2 are relatively prime, then $(n_1, k) = 1$ and the resulting arc joining a to b includes all of the pieces crossing the belt and we have constructed an arc with signature n_1, n_2. If $(n_1, n_2) = q > 1$, then since $2k = n_1 - n_2$, $(n_1, k) = q$ also, and the curve system will contain an arc joining a to b with signature n_1/p, n_2/q and one or more closed curves. So, if $(n_1, n_2) = 1$, there exists an arc with signature n_1, n_2 while if we start with a *standard* system of arcs connected directly across as described above with signature $1, m$ ($m < 0$), then no set of Dehn Twists will result in an arc with a signature n_1, n_2 for which $(n_1, n_2) > 1$.

Next, suppose that there exists an arc joining a to b which has signature n_1, n_2 with $n_1 > 0$, $n_1 \geq n_2$ and $(n_1, n_2) > 1$. Do a Dehn Twist of $-j$ where $2j = n_1 - 1$. This results in an arc system having signature $1, m$ with $m \leq 1$ and containing an arc joining a to b. If $m = 1$, then a and b are connected directly and there can be no arcs separating a and b. Thus there are no crossings of the belt and Dehn twists will have no effect. That is, there are no homotopy classes with signature n_1, n_2, $n_1 > 1$.

It remains only to show that the only possible system with signature $1, m$, $m > 0$, is one of the type described above with direct connections across the belt joining a system of k arcs on the left to the corresponding standard system on the left. The first integer in the signature being 1 requires that the connections across the belt be direct or direct plus a twist of ± 1 since otherwise there will be more than one arc around both a and b. A twist of -1 clearly requires that $m > 0$. A twist of $+1$ would give a first integer > 1 also, since there would be 1 from the connection to L_1 and two more from the arc going around the belt to connect L_0 to R_{2k-1}. Thus the only feasible case is the set of direct connections and the theorem is proved. ■

5. Remarks

Every simple curve joining a and b in $\Delta \backslash \{0, a, b\}$ defines a signature n_1, n_2 and every signature defines a θ_0. However, there are only a countable number of such signatures. Hence, almost all θ_0 must give rise to a trajectory structure in which the unit disk is a density domain. How is it that density domains do not seem to give rise to problems in the calculations, particularly since numerical methods are of necessity not exact? The answer is that we are looking for information about the mappping and most of this information is a continuous function of the parameter θ_0. It is important to verify that a θ_0 exists defining the solution to the problem. Then nearby θ_0 will define an approximation to the solution. The exception to this principle occurs when we consider properties that depend on the homotopy class of γ, such as the length of the arc γ. For example, the curves with signature $n, n+2$ with n odd and $n, n+2$ relatively prime define angles θ_0 which converge to the θ_0 corresponding to the signature 1,1, but the lengths of the γ tend to infinity.

The above discussion is necessarily somewhat vague since each problem will be different and one is faced with many difficult topological problems, each which must be solved on its own merits. However, it is clear that the Schiffer Differential Equation can usually be solved numerically with the help of standard numerical integration and multidimensional zero finding techniques.

As a closing remark, we observe that one of the biggest problems in the numerical solution of the Schiffer Differential Equation is in determining the argument of the functions being integrated. For example if we integrate $R(w)^{1/2}dw$ along some path, it is essential that the argument of the integrand be continuous. The complex square root in FORTRAN or other computer languages has a jump if the variable crosses the negative real axis and this can easily lead to erroneous results. The best way of overcoming this problem is to do the computation of individual factors separately, choosing forms which do not cross the negative real axis along the path of integration. For example, the argument of the factor $(b - w)$ in (3) will change by at most π along a line segment, so, if it is multiplied by a constant as necessary so that at least one point along the path makes $(b - w)$ real and positive, there will be no jump in the argument. The alternative is to adjust the argument at each point to make sure that it is continuous.

References

[1] A. Fathi, F. Laudenbach, V. Poenaru, et al., *Travaux de Thurston sur les surfaces*, Asterisque, (1979), 66-67.

[2] P.R. Garabedian, M. Schiffer, *The local maximum theorem for the coefficients of univalent functions*, Arch. Rational Mech. Anal., **26** (1967), 1-32.

[3] J. Hummel, B. Pinchuck, *Variations for bounded nonvanishing univalent functions*, J. Analyse Math., **44** (1984/85), 183-199.

[4] J. Hummel, B. Pinchuk, *A minimal distance problem in conformal mapping*, Complex Variables, **9** (1987), 211-220.

[5] J. Krzyż, *An extremal length problem and its applications*, Proc. of the NRL Conference on Classical Function Theory (1970), Math. Research Center, Naval Research Laboratory, Washington D.C., 143-155.

[6] L. Liao, *Certain extremal problems concerning module and harmonic measure*, J. Analyse Math., **40** (1981), 1-42.

Received: April 5, 1989.

Computational Methods and Function Theory
Proceedings, Valparaíso 1989
St. Ruscheweyh, E.B. Saff, L. C. Salinas, R.S. Varga (*eds.*)
Lecture Notes in Mathematics **1435**, pp. 81–87

Behavior of the Lagrange Interpolants in the Roots of Unity

K.G. Ivanov[1]

Institute of Mathematics, Bulgarian Academy of Science
Sofia, 1090, Bulgaria

and

E.B. Saff[2]

Institute for Constructive Mathematics, Department of Mathematics
University of South Florida, Tampa
Florida 33620, USA

Dedicated to R.S. Varga on the occasion of his sixtieth birthday.

Abstract. Let A_0 be the class of functions f analytic in the open unit disk $|z| < 1$, continuous on $|z| \leq 1$, but not analytic on $|z| \leq 1$. We investigate the behavior of the Lagrange polynomial interpolants $L_{n-1}(f, z)$ to f in the $n-$th roots of unity. In contrast with the properties of the partial sums of the Maclaurin expansion, we show that for any w, with $|w| > 1$, there exists a $g \in A_0$ such that $L_{n-1}(g, w) = 0$ for all n. We also analyze the size of the coefficients of $L_{n-1}(f, z)$ and the asymptotic behavior of the zeros of the $L_{n-1}(f, z)$.

1. Convergence

Let $f(z) = \sum_{k=0}^{\infty} a_k z^k$ be continuous on $D_1 := \{z \in \mathbb{C} : |z| \leq 1\}$. Then the Lagrange interpolant to f at the $n-$th roots of unity $e(\frac{k}{n}), k = 0, 1, \ldots, n-1, e(x) := e^{2\pi i x}$, can be written as

[1]The research of this author was conducted while visiting the University of South Florida.
[2]The research of this author was supported, in part, by the National Science Foundation under grant DMS-881-4026.

$$L_{n-1}(f,z) = \frac{z^n - 1}{n} \sum_{k=0}^{n-1} \frac{e(\frac{k}{n}) f(e(\frac{k}{n}))}{z - e(\frac{k}{n})} =: \sum_{j=0}^{n-1} c(j,n) z^j, \tag{1.1}$$

where

$$c(j,n) := \frac{1}{n} \sum_{k=0}^{n-1} e\left(\frac{(n-j)k}{n}\right) f\left(e(\frac{k}{n})\right), \; j = 0, 1, \ldots, n-1. \tag{1.2}$$

When f is analytic on D_1 (that is, f is analytic on $|z| \leq 1+\epsilon$ for some $\epsilon > 0$), several results concerning Walsh's theory of equiconvergence describe the very close behavior of the sequence of Lagrange interpolants $\{L_{n-1}(f,z)\}$ and the sequence of partial sums $\{s_{n-1}(f,z)\}$ of its Taylor series, $s_{n-1}(f,z) := \sum_{k=0}^{n-1} a_k z^k$. For example, for such f's and for any $z \in \mathbb{C}$, the sequences $\{L_{n-1}(f,z)\}_1^\infty$ and $\{s_{n-1}(f,z)\}_1^\infty$ either both converge or both diverge (hence the term *equiconvergence*).

But when f belongs to A_0 — the set of all functions continuous on D_1, analytic on the interior $\mathring{D}_1$, but not analytic on D_1 —, the behavior of the two sequences may be different. Of course, both $\{L_{n-1}(f,z)\}_1^\infty$ and $\{s_{n-1}(f,z)\}_1^\infty$ converge (to $f(z)$) when $|z| < 1$. When $|z| = 1$ there are several examples of $f \in A_0$ such that $\{L_{n-1}(f,z)\}_1^\infty$ converges but $\{s_{n-1}(f,z)\}_1^\infty$ diverges (the first goes back to du Bois-Reymond who constructed a function $f \in A_0$ with a divergent Maclaurin series at $z = 1$, but $L_{n-1}(f,1) = f(1)$). Conversely, $\{L_{n-1}(f,z)\}_1^\infty$ may diverge at a point on $|z| = 1$ where $\{s_{n-1}(f,z)\}_1^\infty$ converges (if f is continuous and of bounded variation on $|z| = 1$, then $s_{n-1}(f)$ converges *uniformly* to f, but $L_{n-1}(f,z)$ can diverge for appropriate f and z, e.g. $f(z) = (1 - \log(1+z))^{-1/2}$ at $z = -1$). When $|z| > 1$, then $\{s_{n-1}(f,z)\}_1^\infty$ necessarily diverges (the terms $a_n z^n$ do not tend to zero). Surprisingly, it is still possible for $\{L_{n-1}(f,z)\}_1^\infty$ to converge for some z with $|z| > 1$, as the corollary of the following theorem shows.

Theorem 1. *Let Λ be any subset of $\mathbb{N}$ and let $m \in \mathbb{N}$. The following are equivalent:*

(a) *There exists an $f \in A_0$ such that the first m coefficients $c(j,n), j = n-1, \ldots, n-m$ of $L_{n-1}(f,z)$ are zero for every $n \in \Lambda$.*

(b) *There exist distinct points $w_j, |w_j| > 1, j = 1, 2, \ldots, m$, and $g \in A_0$ such that $L_{n-1}(g, w_j) = 0$ for every $j = 1, 2, \ldots, m$ and every $n \in \Lambda$.*

Proof. **(a) ⇒ (b).** From (1.2) we have

$$\sum_{k=0}^{n-1} e(\frac{ks}{n}) f(e(\frac{k}{n})) = 0, \quad s = 1, 2, \ldots, m, \quad n \in \Lambda. \tag{1.3}$$

Let $w_j, j = 1, \ldots, m$, be any m different points in $|z| > 1$ and let $g(z) := f(z)p(z)$, where $p(z) := \prod_{j=1}^m (z - w_j)$. Setting

$$W_{s,j} := (-1)^s \sum_{\substack{\Sigma\alpha_\nu = s \\ \alpha_j = 0 \\ \alpha_\nu = 0 \text{ or } 1}} w_1^{\alpha_1} \ldots w_m^{\alpha_m}$$

and using (1.1) and (1.3) we get that for any $n \in \Lambda$ and $j = 1, 2, \ldots, m$

$$
\begin{aligned}
L_{n-1}(g,w_j) &= \frac{w_j^n-1}{n}\sum_{k=0}^{n-1} f(e(\frac{k}{n}))e(\frac{k}{n})\frac{p(e(\frac{k}{n}))}{w_j-e(\frac{k}{n})} \\
&= -\frac{w_j^n-1}{n}\sum_{k=0}^{n-1} f(e(\frac{k}{n}))e(\frac{k}{n})\prod_{\substack{\nu=1\\ \nu\neq j}}^{m}(e(\frac{k}{n})-w_\nu) \\
&= -\frac{w_j^n-1}{n}\sum_{k=0}^{n-1} f(e(\frac{k}{n}))e(\frac{k}{n})\sum_{s=1}^{m} e\left(\frac{k(s-1)}{n}\right) W_{m-s,j} \\
&= -\frac{w_j^n-1}{n}\sum_{s=1}^{m} W_{m-s,j}\sum_{k=0}^{n-1} f(e(\frac{k}{n}))e(\frac{ks}{n}) = 0.
\end{aligned}
\tag{1.4}
$$

(b) ⇒ (a). Keeping the notation from the first part of the proof we again set $f(z) := g(z)/p(z)$. Then $f \in A_0$ because the w_j's are outside D_1. From (1.4) we have

$$
\sum_{s=1}^{m} W_{m-s,j}c(n-s,n) = 0, \quad j = 1,2,\ldots,m,
$$

for any $n \in \Lambda$. But $\mathrm{Det}(W_{m-s,j})_{j=1,s=1}^{m\ \ m} = \Pi_{k<l}(w_k - w_l) \neq 0$. Hence $c(n-s,n) = 0$, $s = 1,\ldots,m$, which completes the proof. ∎

Corollary 2. *For any* $w, |w| > 1$, *there is a* $g \in A_0$ *such that* $L_{n-1}(g,w) = 0$ *for every* $n \in \mathbb{N}$.

Proof. According to Theorem 1 it is enough to find $f \in A_0$ for which all leading coefficients of the Lagrange interpolants are zero. For the function

$$
F(z) := \sum_{k=1}^{\infty} \frac{\mu(k)}{k} z^k,
$$

where μ is the Möbius function (of number theory), we know (see [1],[5]) that

$$
\sum_{k=0}^{n-1} F(e(\frac{k}{n})) = 0, \quad n \in \mathbb{N}.
$$

Hence for $f(z) := F(z)/z$ we have $\Sigma_{k=0}^{n-1} f(e(\frac{k}{n}))e(\frac{k}{n}) = 0$ for any n, which proves the corollary. In this case, $g(z) = \Sigma_{k=1}^{\infty}\left(\frac{\mu(k)}{k} - w\frac{\mu(k+1)}{k+1}\right) z^k - w$. ∎

Remark 1. We do not know whether there exists a $g \in A_0$ such that $L_{n-1}(g,w_j) = 0$, $j = 1,2$, for every $n \in \mathbb{N}$, where $|w_j| > 1, w_1 \neq w_2$.

Remark 2. Any g satisfying Theorem 1 or Corollary 2 will not be smooth. For example, no function with absolutely convergent Maclaurin series on $|z| = 1$ can satisfy Corollary 2.

2. Coefficients and the Distribution of Zeros

According to a theorem of Jentzsch [6], for any function $f \in A_0$, every point on the boundary of D_1 is a limit point of the zeros of $\{s_{n-1}(f,z)\}_1^\infty$. One can say even more — the zeros of a special subsequence $\{s_{n_j-1}(f,z)\}$ tend weakly to the uniform distribution on the unit circle $\{z : |z| = 1\}$ (see Szegö [7]). The same behavior can be observed for the zeros of the best polynomial approximants to $f \in A_0$ (see [2,3]). It is natural to ask whether the sequence of Lagrange interpolants $\{L_{n-1}(f,z)\}_1^\infty$ also possesses this property.

A crucial step in establishing the above mentioned facts is the proof that the leading coefficients of the full sequence of polynomials are not geometrically small. For example, for the partial sums of Taylor series, this means

$$\limsup_{n\to\infty} |a_n|^{1/n} = 1,$$

which is equivalent to $f \in A_0$, provided $f \in C(D_1)$.

One cannot expect the same behavior for the leading coefficients of $L_{n-1}(f)$. Indeed, as the example function $f \in A_0$ from the proof of Corollary 2 shows, we may have $c(n-1,n) = 0$ for every n. But results similar to Jentzsch's and Szegö's theorems still can be established by utilizing the following statement, which is a special case of Theorem 1 in Grothmann [4].

Let p_m be an algebraic polynomial of exact degree $\kappa(m)$. Define the *zero-measure* ν_m associated with p_m as

$$\nu_m(A) := \frac{\#\text{ of zeros of } p_m \text{ in } A}{\kappa(m)}$$

for any Borel set $A \subset \mathbb{C}$, where the zeros are counted with their multiplicity.

Theorem A. ([4]) *Let Λ be a sequence of positive integers and assume that the following three conditions hold for the sequence $\{p_m\}_{m\in\Lambda}$ of algebraic polynomials:*

(i) $\limsup\limits_{\substack{m\to\infty\\ m\in\Lambda}} \left(\sup\limits_{z\in D_1} \frac{1}{\kappa(m)} \log|p_m(z)| \right) \le 0$;

(ii) *for every compact set* $M \subset \mathring{D}_1$,

$$\lim_{\substack{m\to\infty\\ m\in\Lambda}} \nu_m(M) = 0;$$

(iii) *there is a compact set* $K \subseteq \overline{\mathbb{C}} \backslash D_1$ *with*

$$\liminf_{\substack{m\to\infty\\ m\in\Lambda}} \left[\sup_{z\in K} \left(\frac{1}{\kappa(m)} \log|p_m(z)| - \log|z| \right) \right] \ge 0.$$

Then, in the weak-star topology, ν_m tends to the uniform distribution $\frac{1}{2\pi}d\theta$ on the unit circle as $m \to \infty, m \in \Lambda$.

This leads us to investigating

$$\sigma(f,\theta) := \limsup_{n\to\infty} \max_{(1-\theta)n\le j<n} |c(j,n)|^{1/n}$$

for $\theta \in (0,1]$, where $c(j,n)$ is defined in (1.2). Obviously $\sigma(f,1) = 1$ and σ is an increasing function of θ for any $f \in A_0$. We shall prove

Theorem 3. *For any $f \in A_0$, we have $\sigma(f,1/3) = 1$.*

Proof. For any $r \in \mathbf{N}$ using (1.2) we get $(0 \le j < n)$

$$\begin{aligned} \textstyle\sum_{l=0}^{r-1} c(j+ln,rn) &= \frac{1}{rn}\sum_{k=0}^{rn-1} f(e(\frac{k}{rn}))\sum_{l=0}^{r-1} e(\frac{-k(j+ln)}{rn}) \\ &= \frac{1}{rn}\sum_{k=0}^{rn-1} f(e(\frac{k}{rn}))e(\frac{-kj}{rn})\sum_{l=0}^{r-1} e(-\frac{kl}{r}) \qquad (2.1)\\ &= \frac{1}{rn}\sum_{m=0}^{n-1} f(e(\frac{m}{n}))e(\frac{-jm}{n})r = c(j,n). \end{aligned}$$

Let us assume that $\sigma(f,1/3) < 1$. Fix q, such that $\sigma(f,1/3) < q < 1$. Then for any $n > n_0$ and any j, $\frac{2}{3}n \le j < n$, we have

$$|c(j,n)| < q^n. \tag{2.2}$$

Fix $l > n_0$ such that $\frac{5}{1-q}q^s < \frac{1}{2}q^l, s = [\frac{3l}{2}]$, and $|a_l| > q^l$ ($f \in A_0$ implies that $\limsup_{n\to\infty}|a_n|^{1/n} = 1$). From the continuity of f and (1.2) we get

$$\lim_{n\to\infty} c(l,n) = \frac{1}{2\pi i}\int_{|z|=1} f(z)z^{-l-1}dz = a_l.$$

Therefore $|c(l,n)| > \frac{1}{2}q^l$ for any $n \ge n_1$. Let us fix $m \in \mathbf{N}$ such that $2\cdot 3^m \cdot 5 \ge n_1$. Then

$$|c(l,2\cdot 3^m\cdot 5)| > \frac{1}{2}q^l. \tag{2.3}$$

Now our aim is, using (2.1) and (2.2), to obtain an estimate contradicting (2.3). From (2.1) we get

$$c(l,2s) = c(l,s) - c(l+s,2s), \tag{2.4}$$

$$c(l,6n) + c(l+2n,6n) + c(l+4n,6n) = c(l,2n), \tag{2.5}$$

and

$$c(l+2n,6n) + c(l+5n,6n) = c(l+2n,3n). \tag{2.6}$$

From (2.5) and (2.6) we get

$$(2.7) \qquad c(l,6n) = c(l,2n) + \{c(l+5n,6n) - c(l+4n,6n) - c(l+2n,3n)\}.$$

From (2.7) with $n = 3^{k-1}s, k = 1,2,\ldots,m$, and (2.4) we obtain

$$(2.8) \qquad c(l, 2\cdot 3^m s) = c(l,s) - c(l+s,2s) + \alpha$$

where

$$\alpha = \sum_{k=1}^{m} \{c(l+5\cdot 3^{k-1}s, 2\cdot 3^k s) - c(l+4\cdot 3^{k-1}s, 2\cdot 3^k s) - c(l+2\cdot 3^{k-1}s, 3^k s)\}.$$

It is easy to see that all terms on the right-hand side of (2.8) are of the type $c(j,n)$ with $\frac{2}{3}n \le j < n$, $n \ge s > l > n_0$. By applying (2.2) in (2.8) we get

$$\begin{aligned} |c(l, 2\cdot 3^m s)| &< q^s + q^{2s} + \textstyle\sum_{k=1}^m \{2q^{2(3^k s)} + q^{3^k s}\} \\ &< q^s + q^{2s} + 3q^{3s}/(1-q) < 5q^s/(1-q) < \frac{1}{2}q^l. \end{aligned}$$

This estimate contradicts (2.3) and proves the theorem. ∎

If one assumes that $f \in A_0$ has an absolutely convergent Maclaurin series on $|z| = 1$, then $\limsup_{n\to\infty} |c(n-1,n)|^{1/n} = 1$ (cf. [5]). This implies that $\sigma(f,\theta) = 1, \theta \in (0,1]$, for such f's.

Theorem 3 and the above observation give some evidence to the following.

Conjecture. *For any $f \in A_0$ and any $0 < \theta < 1$, we have $\sigma(f,\theta) = 1$.*

Now we can establish

Theorem 4. *If the above conjecture is true, then for any $f \in A_0$ there is a subsequence $\{n_j\}$ such that the zero measures ν_{n_j} (corresponding to $L_{n_j-1}(f)$) tend (in the weak-star topology) to the uniform distribution on the unit circle as $j \to \infty$.*

Proof. We are going to apply Theorem A with $\{p_m\}$ an appropriate subsequence $L_{n_j-1}(f)$ of the Lagrange interpolants. The subsequence $L_{n_j-1}(f)$ is chosen so that condition (iii) is satisfied.

Let $r > 1$ be fixed. Assume that there exists an ϵ, $0 < \epsilon < 1/2$, such that

$$\limsup_{n\to\infty} \left[\sup_{|z|=r} \left(\frac{1}{\kappa(n)} \log |L_{n-1}(f,z)| - \log r \right) \right] < -3\epsilon \log r,$$

where $\kappa(n)$ is the precise degree of $L_{n-1}(f,z)$. Then for $n > N_0$ and for every $z, |z| = r$, we have

$$\frac{1}{\kappa(n)} \log |L_{n-1}(f,z)| - \log r < -2\epsilon \log r,$$

that is,

$$|L_{n-1}(f,z)| < r^{(1-2\epsilon)\kappa(n)} \le r^{(1-2\epsilon)n}.$$

Hence for any $j, (1-\epsilon)n \le j < n$, we have

$$|c(j,n)| = \left| \frac{1}{2\pi i} \int_{|z|=r} L_{n-1}(f,z) z^{-j-1} dz \right| \leq r^{(1-2\epsilon)n-j} \leq r^{-\epsilon n}.$$

This inequality implies that $\sigma(f,\epsilon) \leq r^{-\epsilon}$, which contradicts the conjecture. Therefore

$$\limsup_{n\to\infty} \left[\sup_{|z|=r} \left(\frac{1}{\kappa(n)} \log |L_{n-1}(f,z)| - \log r \right) \right] \geq 0.$$

Hence there exists a subsequence $\{n_j\}$ such that

$$\liminf_{j\to\infty} \left[\sup_{|z|=r} \left(\frac{1}{\kappa(n_j)} \log |L_{n_j-1}(f,z)| - \log r \right) \right] \geq 0.$$

Consequently, condition (iii) of Theorem A is fulfilled for the sequence $\{L_{n_j-1}(f)\}$. The other two conditions of Theorem A are easily seen to be satisfied (even for the whole sequence of Lagrange interpolants). Indeed, condition (i) follows from the trivial estimate

$$\| L_{n-1}(f) \|_{D_1} \leq n \| f \|_{D_1}$$

(see (1.1)) and $\kappa(n) \geq \frac{2}{3}n$ (see Theorem 3 — we do not need the Conjecture here). Condition (ii) follows from the facts that $\{L_{n-1}(f)\}$ approximates f uniformly on any compact set $M \subset \overset{\circ}{D}_1$ and that f can have only finitely many zeros on M. Hence Theorem 4 follows from Theorem A. ■

References

[1] G.R. Blakley, I. Borosh, C.K. Chui, *A two dimensional mean problem*, J. Approx. Theory **22** (1978) 11-26.

[2] H.-P. Blatt and E.B. Saff, *Behavior of zeros of polynomials of near best approximation*, J. Approx. Theory **46** (1986), 323-344.

[3] H.-P. Blatt, E.B. Saff, M. Simkani, *Jentzsch-Szegö type theorems for best approximants*, Journ. London Math. Soc. (2) **38** (1988) 307-316.

[4] R. Grothmann, *On the zeros of sequences of polynomials*, J. Approx. Theory (to appear).

[5] K.G. Ivanov, T.J. Rivlin, E.B. Saff, *The representation of functions in terms of their divided differences at Chebyshev nodes and roots of unity*, (to appear).

[6] R. Jentzsch, "Untersuchungen zur Theorie Analytischer Funktionen", Inaugural-dissertation, Berlin, 1914.

[7] G. Szegö, *Über die Nullstellen von Polynomen, die in einem Kreis gleichmäßig konvergieren*, Sitzungsber. Berl. Math. Ges. **21** (1922) 53-64.

Received: October 10, 1989

Computational Methods and Function Theory
Proceedings, Valparaíso 1989
St. Ruscheweyh, E.B. Saff, L. C. Salinas, R.S. Varga (*eds.*)
Lecture Notes in Mathematics **1435**, pp. 89–101

Orthogonal Polynomials, Chain Sequences, Three-term Recurrence Relations and Continued Fractions

Lisa Jacobsen

Division of Mathematical Sciences, The University of Trondheim
N-7034 Trondheim, Norway

1. Introduction

Let Π denote the set of all (complex) polynomials, and let $\mathcal{L} : \Pi \to \mathbb{C}$ be a linear functional. We say that $\{P_n(z)\}_{n=0}^{\infty}$; $P_n \in \Pi$ is a *sequence of orthogonal polynomials* (OPS) with respect to $\mathcal{L}$ iff

(i)	$\deg P_n = n$	for $n = 0, 1, 2, \ldots,$
(ii)	$\mathcal{L}[P_n \cdot P_m] = 0$	if $n \neq m$, and
(iii)	$\mathcal{L}[P_n^2] \neq 0$	for $n = 0, 1, 2, \ldots$.

Such a sequence does not always exist. Indeed, it is simple to see that $\mathcal{L}$ permits an OPS iff

$$(1.1) \qquad \Delta_n := \begin{vmatrix} \mu_0 & \mu_1 & \cdots & \mu_n \\ \mu_1 & \mu_2 & \cdots & \mu_{n+1} \\ \cdot & \cdot & \cdots & \cdot \\ \cdot & \cdot & \cdots & \cdot \\ \cdot & \cdot & \cdots & \cdot \\ \mu_n & \mu_{n+1} & \cdots & \mu_{2n} \end{vmatrix} \neq 0 \quad \text{for } n = 0, 1, 2, \ldots ,$$

where $\{\mu_k\}$ are the *moments* $\mu_k = \mathcal{L}[z^k]$. On the other hand, if $\mathcal{L}$ permits an OPS $\{P_n(z)\}$, then each $P_n(z)$ is unique up to a constant factor. So, in the following we may assume that all P_n are normalized to be monic; i.e. to have leading coefficients 1.

Letting

$$(1.2)\qquad \lambda_{n+1} := \frac{\Delta_{n-2}\Delta_n}{\Delta_{n-1}^2}, \quad c_n := \frac{\Delta_{n-2}}{\Delta_{n-1}}\mathcal{L}[zP_{n-1}^2(z)]; \quad n = 1,2,3,\ldots$$

where $\Delta_{-1} := 1$, one can also show that if $\{P_n(z)\}$ exists, then it is a unique solution of the *three-term recurrence relation*

$$(1.3)\qquad P_n(z) = (z - c_n)P_{n-1}(z) - \lambda_n P_{n-2}(z); \quad \lambda_n \neq 0 \quad \text{for } n = 1,2,3,\ldots$$

where $P_{-1}(z) := 0$, $P_0(z) := 1$ and λ_1 is arbitrary. It is often practical to choose $\lambda_1 := -\mu_0$. By the so-called Favard's theorem, [2, Theorem 4.4, p. 21] one also has that a sequence of polynomials satisfying some three-term recurrence relation (1.3) with the said initial values, is an OPS for some linear functional $\mathcal{L}$. For fuller information on orthogonal polynomials we refer to Chapter 1 in Chihara's book [2].

Associated with a three-term recurrence relation

$$(1.4)\qquad X_n = b_n X_{n-1} + a_n X_{n-2}; \quad a_n \neq 0, \quad a_n, b_n \in \mathbb{C} \quad \text{for } n = 1,2,3,\ldots$$

there is always a *continued fraction*

$$(1.5)\qquad \mathrm{K}\frac{a_n}{b_n} := \frac{a_1}{b_1} + \frac{a_2}{b_2} + \frac{a_3}{b_3} + \ldots + \; = \cfrac{a_1}{b_1 + \cfrac{a_2}{b_2 + \cfrac{a_3}{b_3 + \ldots}}}; \quad a_n \neq 0.$$

For one thing one has (by induction) that its *modified approximants*

$$(1.6)\qquad S_n(w) := \frac{a_1}{b_1} + \frac{a_2}{b_2} + \ldots + \frac{a_n}{b_n + w} = \frac{A_n + A_{n-1}w}{B_n + B_{n-1}w}; \quad n \in \mathbb{N}$$

has coefficients $\{A_n\}$ and $\{B_n\}$ which are solutions of (1.4) with initial values $A_{-1} = 1$, $A_0 = 0$ and $B_{-1} = 0$, $B_0 = 1$. $\{A_n\}$ and $\{B_n\}$ are called the canonical numerators and denominators of $\mathrm{K}(a_n/b_n)$. Moreover, if $\{X_n\}$ is a non-trivial solution of (1.4); $X_n \in \mathbb{C}$, then $\{X_n/X_{n-1}\}$ is well defined in $\hat{\mathbb{C}} = \mathbb{C} \cup \{\infty\}$, and one has

$$(1.7)\qquad -\frac{X_0}{X_{-1}} = \cfrac{a_1}{b_1 - \cfrac{X_1}{X_0}} = \frac{a_1}{b_1} + \cfrac{a_2}{b_2 - \cfrac{X_2}{X_1}} = \ldots = \frac{a_1}{b_1} + \frac{a_2}{b_2} + \ldots + \cfrac{a_n}{b_n - \cfrac{X_n}{X_{n-1}}}.$$

The sequence $\{-X_n/X_{n-1}\}$ is called a *tail sequence* for $\mathrm{K}(a_n/b_n)$. It satisfies

$$(1.8)\qquad t_0 = S_n(t_n) = \frac{A_n + A_{n-1}t_n}{B_n + B_{n-1}t_n} \quad \text{for all } n \in \mathbb{N}; \quad t_n = -X_n/X_{n-1},$$

and

$$(1.9)\qquad t_{n-1} = s_n(t_n) \quad \text{where } s_n(w) := \frac{a_n}{b_n + w} \quad \text{for all } n \in \mathbb{N}.$$

We say that $\mathrm{K}(a_n/b_n)$ *converges* if its sequence of classical approximants $S_n(0) = A_n/B_n$ converges in $\hat{\mathbb{C}}$. Its value is then $f = \lim A_n/B_n$. Clearly $\mathrm{K}(a_n/b_n)$ converges to f if and only if its n-th tail

$$\text{(1.10)} \qquad \underset{k=1}{\overset{\infty}{\mathrm{K}}} \, (a_{n+k}/b_{n+k}) = \frac{a_{n+1}}{b_{n+1}} \, \frac{}{+} \, \frac{a_{n+2}}{b_{n+2}} \, \frac{}{+} \, \dots$$

converges to $f^{(n)}$ for all n, where $\{f^{(n)}\}$ is a tail-sequence for $\mathrm{K}(a_n/b_n)$ with $f^{(0)} = f$. For more information on continued fractions, see Chapter 2 in the book by Jones and Thron [6].

We see immediately that the OPS $\{P_n(z)\}$ in (1.3) gives the canonical denominators of the continued fraction

$$\text{(1.11)} \qquad \frac{\lambda_1}{z - c_1} \, \frac{}{-} \, \frac{\lambda_2}{z - c_2} \, \frac{}{-} \, \frac{\lambda_3}{z - c_3} \, \frac{}{-} \, \dots \qquad \lambda_n \neq 0,$$

a so-called J-fraction or Jacobi-fraction. Further, the (quasi-orthogonal) polynomials

$$\text{(1.12)} \qquad P_n^*(z, w_n) := P_n(z) + w_n P_{n-1}(z); \quad w_n \in \mathbb{C}$$

are the canonical denominators of $S_n(w_n)$. We shall mainly use this connection to derive information on the location of the zeros of $P_n(z)$ and $P_n^*(z, w_n)$, a very important issue.

In the special case where all $\lambda_n > 0$ and all $c_n \in \mathbb{R}$, one has proved that all the zeros of each $P_n(z)$ are real and simple. To locate an interval containing all zeros of all $P_n(z)$, the concept of *chain sequences* has been a useful tool. $\{\alpha_n\}_{n=1}^{\infty}$ is said to be a chain sequence if

$$\text{(1.13)} \qquad \alpha_n := (1 - g_{n-1})g_n; \quad 0 \leq g_0 < 1, \quad 0 < g_n < 1 \quad \text{for } n = 1, 2, 3, \dots$$

for some such sequence $\{g_n\}$. $\{g_n\}$ is then called a parameter sequence for $\{\alpha_n\}$. At this conference the question was raised by Paul Nevai of whether chain sequences could be generalized to yield information in the complex case ($\lambda_n \in \mathbb{C}$, $c_n \in \mathbb{C}$). In Section 3 we shall see that this is indeed so.

The basis for the arguments to be used is the following simple observation:

$$\text{(1.14)} \qquad S_n(w) = \infty \Leftrightarrow \begin{cases} B_n + B_{n-1}w = 0 & \text{or} \\ w = \infty \text{ and } B_{n-1} = 0. \end{cases}$$

(*Proof.* A_n and B_n cannot both be zero since by induction on n one has

$$\text{(1.15)} \qquad A_n B_{n-1} - B_n A_{n-1} = (-1)^{n-1} \prod_{j=1}^{n} a_j \neq 0.$$

Further, $A_n + A_{n-1}w$ and $B_n + B_{n-1}w$ cannot both be zero since

$$\text{(1.16)} \qquad (A_n + A_{n-1}w)B_{n-1} - (B_n + B_{n-1}w)A_{n-1} = A_n B_{n-1} - B_n A_{n-1} \neq 0. \blacksquare)$$

This observation (1.14) means that locating zero-free sets for $P_n(z)$ (or $P_n^*(z, w_n)$; $w_n \neq \infty$) is equivalent to finding sets where $S_n(0)$ (or $S_n(w_n)$) is finite. This can again be done by means of element sets and various types of value sets for continued fractions. These are explained in Section 2.

Finally, the close connection between tail sequences for continued fractions and parameter sequences for chain sequences is discussed in Section 4.

2. Element and value sets for continued fractions

Following [6] we say that $\{\Omega_n\}_{n=1}^{\infty}$ is a *sequence of element sets* for a continued fraction $\mathrm{K}(a_n/b_n)$ iff $(a_n, b_n) \in \Omega_n \subseteq \mathbb{C}^2$ for all n. Even though each Ω_n does not have to contain more than the one point (a_n, b_n), it is often of advantage to consider larger sets Ω_n. Further, we say that $\{V_n\}_{n=0}^{\infty}$ is a *sequence of pre value sets* corresponding to $\{\Omega_n\}$ iff $\emptyset \neq V_n \subseteq \hat{\mathbb{C}}$ and

$$(2.1) \qquad (a_n, b_n) \in \Omega_n \Rightarrow \frac{a_n}{b_n + V_n} \subseteq V_{n-1}; \quad \forall n \in \mathbb{N}.$$

The importance of these sets lies in the fact that if $(a_k, b_k) \in \Omega_k$ for all k and $w_n \in V_n$, then

$$(2.2) \qquad S_n(w_n) := \frac{a_1}{b_1} + \frac{a_2}{b_2} + \ldots + \frac{a_n}{b_n + w_n} = \frac{A_n + A_{n-1}w_n}{B_n + B_{n-1}w_n} \in V_0.$$

So, if $w_n \neq \infty$ and $\infty \notin V_0$, then $B_n + B_{n-1}w_n \neq 0$. If in particular $0 \in V_n$, or more generally

$$(2.3) \qquad (a_n, b_n) \in \Omega_n \Rightarrow a_n/b_n \in V_{n-1},$$

then $A_n/B_n \in V_0$, so that $\infty \notin V_0 \Rightarrow B_n \neq 0$. If (2.3) holds for all n, we say that $\{V_n\}$ is a *sequence of value sets* corresponding to $\{\Omega_n\}$.

Still another type of value sets is the following: $\{W_n\}_{n=0}^{\infty}$ is called a *sequence of value sets for tails* corresponding to $\{\Omega_n\}$ iff

$$(2.4) \qquad (a_n, b_n) \in \Omega_n \Rightarrow -b_n + \frac{a_n}{W_{n-1}} \subseteq W_n; \ \forall n \in \mathbb{N}.$$

This means that if $\{t_n\}$ is a tail sequence for $\mathrm{K}(a_n/b_n)$ with $t_{n-1} \in W_{n-1}$, then $t_n \in W_n$ and thus $t_k \in W_k$ for all $k \geq n$. By (1.7) it follows that $\{-B_n/B_{n-1}\}$ is a tail sequence for $\mathrm{K}(a_n/b_n)$. Since by (1.15) B_n and B_{n-1} cannot both be zero, it follows that $B_{n-1} \neq 0$ if $-B_n/B_{n-1} \in W_n$ and $\infty \notin W_n$. Likewise, $B_n \neq 0$ if $-B_n/B_{n-1} \in W_n$ and $0 \notin W_n$.

Finding pre value sets $\{V_n\}$ and finding value sets for tails $\{W_n\}$ corresponding to a sequence $\{\Omega_n\}$ of element sets, is really the same thing. Indeed, $\{V_n\}$; $V_n \neq \emptyset, \hat{\mathbb{C}}$ for all n, is a sequence of pre value sets for $\{\Omega_n\}$ if and only if $\{\hat{\mathbb{C}} \setminus V_n\}$ is a sequence of value sets for tails for $\{\Omega_n\}$. (*Proof.* $s_n(w) := a_n/(b_n + w)$ is a non-singular linear fractional transformation, and thus a bijective mapping of $\hat{\mathbb{C}}$ onto $\hat{\mathbb{C}}$. Hence $s_n(V_n) \subseteq V_{n-1} \Leftrightarrow V_n \subseteq s_n^{-1}(V_{n-1}) \Leftrightarrow \hat{\mathbb{C}} \setminus V_n \supseteq s_n^{-1}(\hat{\mathbb{C}} \setminus V_{n-1})$.■)

Another matter is that starting with element sets $\{\Omega_n\}$ and trying to find corresponding pre value sets $\{V_n\}$ is tough going. The "right" thing to do is to cheat, and start with $\{V_n\}$ or $\{W_n\}$ and then find a sequence $\{\Omega_n\}$ of element sets to which they correspond. Since we are actually mapping V_n or W_n by means of linear fractional transformations s_n or s_n^{-1}, it is natural to let V_n of W_n be halfplanes, interiors of circles or exterior of circles. Let us look at some well known examples.

Example 2.1. Let

$$V_n := \{z \in \mathbb{C}; |z| \leq (1-g_n)\} \quad \text{for } n = 0,1,2,\ldots$$

where $0 \leq g_0 < 1$, $0 < g_n < 1$ for $n \in \mathbb{N}$. Then $\{V_n\}$ is a sequence of value regions for $\{\Omega_n\}$ where $\Omega_n := E_n \times \{1\}$ and

$$E_n := \{z \in \mathbb{C};\ |z| \leq (1-g_{n-1})g_n\}.$$

This example originally dates back to Worpitzky 1865, [13], but the present form is due to Wall [12, Theorem 11.1, p. 45-46]. Clearly, the radii of E_n form a chain sequence.

Since all Ω_n have the form $E_n \times \{1\}$, $\{\Omega_n\}$ is a sequence of element sets for continued fractions $\mathrm{K}(a_n/1)$; $a_n \in E_n$. Every continued fraction $\mathrm{K}(a_n/b_n)$ with all $b_n \neq 0$ can be brought to the form $\mathrm{K}(d_n/1)$ by the equivalence transformation

$$\mathrm{K}(a_n/b_n) \approx \mathrm{K}(d_n/1) \Leftrightarrow d_n = a_n/b_n b_{n-1},\ b_0 := 1. \tag{2.5}$$

The connection between $\mathrm{K}(a_n/b_n)$ and $\mathrm{K}(d_n/1)$ is then that

$$S_n(w) := \frac{a_1}{b_1} + \frac{a_2}{b_2} + \ldots + \frac{a_n}{b_n + w} = \frac{d_1}{1} + \frac{d_2}{1} + \ldots + \frac{d_n}{1 + w/b_n} =: T_n(w/b_n), \tag{2.6}$$

so that $S_n(w) = \infty \Leftrightarrow T_n(w/b_n) = \infty$.

The next example is part of the most general version for the celebrated parabola theorem. It is due to Jones and Thron, [5],[6, Theorem 4.4, p. 68]. (The first parabola theorem dates back to Scott and Wall [10], 1940).

Example 2.2. Let

$$V_n := \{z \in \mathbb{C};\ \mathrm{Re}(ze^{-i\psi_n}) \geq -p_n\};\ p_n > 0,\ \psi_n \in \mathbb{R} \quad \text{for } n = 0,1,2,\ldots \tag{2.7}$$

Then $\{V_n\}$ is a sequence of value regions corresponding to

$$\Omega_n := \{(a_n, b_n) \in \mathbb{C}^2;\ |a_n| - \mathrm{Re}(a_n e^{-i(\psi_n + \psi_{n-1})}) \leq 2p_{n-1}(\mathrm{Re}(b_n e^{-i\psi_n}) - p_n)\} \tag{2.8}$$

for $n = 1,2,3,\ldots$. It is called a parabola theorem since for fixed b_n, $(a_n, b_n) \in \Omega_n$ if and only if a_n is contained in some region bounded by a parabola. A more familar form of the parabola theorem is obtained if we choose all $\psi_n = \psi$, and require all b_n to be 1:

Let ψ be a fixed real number such that $-\pi/2 < \psi < \pi/2$, and let

$$V_n := \{z \in \mathbb{C};\ \mathrm{Re}(ze^{-i\psi}) \geq -(1-g_n)\cos\psi\} \quad \text{for } n = 0,1,2,\ldots \tag{2.9}$$

where $0 \leq g_0 < 1$ and $0 < g_n < 1$ for $n \in \mathbb{N}$. Then $\{V_n\}$ is a sequence of value regions corresponding to $\{\Omega_n\}$ given by $\Omega_n := E_n \times \{1\}$ where

$$E_n := \{z \in \mathbb{C};\ |z| - \mathrm{Re}(ze^{-2i\psi}) \leq 2(1-g_{n-1})g_n \cos^2\psi\}. \tag{2.10}$$

Here, p_n in (2.8) is replaced by $(1-g_n)\cos\psi$ to make it easier to recognize the chain sequence $(1-g_{n-1})g_n$ in the definition of E_n.

Note that the half planes V_n do not contain ∞.

Example 2.3. Let

$$V_n := \{z \in \mathbb{C};\ |z - \Gamma_n| \le \rho_n\};\ \Gamma_n \in \mathbb{C},\ \rho_n > 0 \quad \text{for } n = 0, 1, 2, \dots . \tag{2.11}$$

Then $\{V_n\}$ is a sequence of pre value regions corresponding to

$$\Omega_n := \{(a_n, b_n) \in \mathbb{C}^2;\ |a_n(\overline{b_n + \Gamma_n}) - \Gamma_{n-1}(|b_n + \Gamma_n|^2 - \rho_n^2)| + |a_n|\rho_n \le \rho_{n-1}(|b_n + \Gamma_n|^2 - \rho_n^2)\} \quad \text{for } n = 1, 2, 3, \dots . \tag{2.12}$$

This was essentially proved by Lane, [7], [6, Theorem 4.3, p. 67], although he was interested only in cases where $0 \in V_n$. This result has in later years been referred to as (part of) the oval theorem, since for fixed b_n, $(a_n, b_n) \in \Omega_n$ if and only if a_n is contained in a domain bounded by a cartesian oval.

If we require that all $b_n = 1$, the element sets $\Omega_n := E_n \times \{1\}$ are given by

$$E_n := \begin{cases} \{z \in \mathbb{C};\quad |z| \le \rho_{n-1}(|1 + \Gamma_n| - \rho_n)\} & \text{if } \Gamma_{n-1} = 0, \\ \{\tilde{a}_n\zeta \in \mathbb{C};\ |\zeta - 1| + \dfrac{\rho_n}{|1 + \Gamma_n|}|\zeta| \le \dfrac{\rho_{n-1}}{|\Gamma_{n-1}|}\} & \text{if } \Gamma_{n-1} \ne 0, \end{cases} \tag{2.13}$$

where

$$\tilde{a}_n = \Gamma_{n-1}(1 + \Gamma_n)(1 - \frac{\rho_n^2}{|1 + \Gamma_n|^2}) \quad \text{for } n = 1, 2, 3, \dots . \tag{2.14}$$

However, in order that $E_n \ne \emptyset$ we need the conditions

$$|1 + \Gamma_n| > \rho_n \quad \text{and} \quad |\Gamma_{n-1}|\rho_n \le |1 + \Gamma_n|\rho_{n-1} \quad \text{for } n = 1, 2, 3, \dots . \tag{2.15}$$

For information on cartesian ovals, see [4]. Let it only be mentioned here that one easily sees that ∂E_n is a cartesian oval with axis along the ray $\arg z = \arg \tilde{a}_n$, foci at $z = 0$ and $z = \tilde{a}_n$, and that $\tilde{a}_n \in E_n$.

For many more examples, see Chapter 4 in Jones' and Thron's book [6].

3. Zerofree regions for orthogonal polynomials

Let us start with a simple case. Let all $\lambda_n > 0$ and all $c_n \in \mathbb{R}$ in (1.3) so that all zeros of $P_n(z)$ are real. Via the equivalence transformation (2.5), the J-fraction (1.11) assumes the form

$$\mathrm{K}(d_n(z)/1) \quad \text{where} \quad d_1(z) := \frac{\lambda_1}{z - c_1},\ d_n(z) := \frac{-\lambda_n}{(z - c_{n-1})(z - c_n)} \tag{3.1}$$

for $n \ge 2$, as long as $z \ne c_n$ for all n. From Example 2.1 and (2.6) it follows then that all $P_n(z) \ne 0$ if $z \ne c_n$ for all n and $|d_n(z)| \le (1 - g_{n-1})g_n$ for all n. It is simple to prove that if $0 < |d_n(z)| \le (1 - g_{n-1})g_n$, where $\{(1 - g_{n-1})g_n\}$ is a chain sequence, then also

$\{|d_n(z)|\}$ is a chain sequence for this particular z. (See for instance [2, Theorem 5.7,p. 97].) So:

Observation 3.1. *All $P_n(z) \neq 0$ for all $z \in \mathbb{C}$ such that $\{|d_n(z)|\}$ is a chain sequence.*

This is of course already well known, [2, Theorem 2.1, p. 108]. Note however that Observation 3.1 also holds if λ_n and c_n are complex numbers. As a very simple corollary we find for instance:

Corollary 3.2. *If M and N are two positive constants such that $|\lambda_n| \leq M^2$ and $|c_n| \leq N$ for all n, then all zeros of all $P_n(z)$ lie in the disk $|z| < 2M + N$.*

Proof. If $|z| \geq 2M + N$, then $|d_n(z)| \leq M^2/(|z| - N)^2 \leq 1/4$. The result then follows since $\{\alpha_n\}_{n=1}^{\infty}$ with all $\alpha_n := 1/4$ is a chain sequence; $1/4 = (1 - 1/2) \cdot \frac{1}{2}$. ■

This is also well known, compare Wall [12, Theorem 26.6, p. 112]. Note also that since $P_n(z)/\prod_{k=1}^{n}(z - c_k)$ are the canonical denominators of $\mathrm{K}(d_n(z)/1)$, we also have that

$$P_n(z) + (z - c_n)w_n(z)P_{n-1}(z) \neq 0 \quad \text{for } n = 1, 2, 3, \ldots$$

for $|z| \geq 2M + N$ as long as $|w_n(z)| \leq 1/2$.

Let us now turn to the parabola theorem (2.9)-(2.10) from Example 2.2. For convenience we exend the definition of chain sequences as follows:

$$\left.\begin{array}{r} \{\alpha_n\} \text{ chain sequence} \\ \\ 0 \leq \tilde{\alpha}_n \leq \alpha_n \text{ for all } n \end{array}\right\} \Rightarrow \{\tilde{\alpha}_n\} \text{ chain sequence.} \tag{3.2}$$

Then we have immediately:

Observation 3.3. *All $P_n(z) \neq 0$ for all $z \in \mathbb{C}$ for which $\{\delta_n(z, \psi)\}_{n=1}^{\infty}$ given by*

$$\delta_n(z, \psi) := \frac{|d_n(z)| - Re(d_n(z)e^{-2i\psi})}{2\cos^2\psi} \tag{3.3}$$

is a chain sequence for some $\psi \in < -\pi/2, \pi/2 >$.

Naturally ψ may vary with z. If we fix ψ to be 0, then this observation still represents a generalization of Observation 3.1. In some cases the condition is easier to check if we write $\delta_n(z, \psi)$ in the form

$$\delta_n(z, \psi) = \left[\frac{\mathrm{Im}(\sqrt{d_n(z)}e^{-i\psi})}{\cos\psi}\right]^2. \tag{3.4}$$

That these two expressions for $\delta_n(z, \psi)$ really are identical is easily checked.

Also from Observation 3.3 one can derive corollaries, such as for instance:

Corollary 3.4. *Let all $c_n = 0$ and all λ_n be real, and let M be a positive constant such that $\lambda_n \leq M^2$ for all n. Then all zeros of all $P_n(z)$ are contained in the strip $|Re z| < 2M$.*

Proof. From Observation 3.3 and (3.4) it follows that all $P_n(z) \neq 0$ for each $z \in \mathbb{C}$ which satisfies

$$\left|\mathrm{Im}\frac{\sqrt{-\lambda_n}}{ze^{i\psi}}\right| \leq \frac{\cos\psi}{2} \quad \text{for all } n \tag{3.5}$$

for some $\psi \in < -\pi/2, \pi/2 >$. By choosing $\psi = -\arg z$ if $|\arg z| < \pi/2$, and $\psi = \pi - \arg z$ or $\psi = -\pi + \arg z$ if $\pi/2 < |\arg z| \leq \pi$ to get $\psi \in < -\pi/2, \pi/2 >$, (3.5) reduces to

$$\left|\mathrm{Im}\sqrt{-\lambda_n}/|z|\right| \leq |\mathrm{Re}z|/2|z| \quad \text{for all } n,$$

which holds if $|\mathrm{Re}z| \geq 2M$. ∎

Again we can also get results for quasi-orthogonal polynomials since

$$P_n(z) + zw_n(z)P_{n-1}(z) \neq 0 \quad \text{for } n = 1, 2, 3, \dots$$

for $|\mathrm{Re}z| \geq 2M$ as long as $\mathrm{Re}(w_n(z)e^{-i\psi}) \geq -\frac{1}{2}\cos\psi$.

Because of the form of the J-fraction (1.11) it is usually simpler to apply the parabola theorem (2.7)-(2.8):

Observation 3.5. *All $P_n(z) \neq 0$ for all $z \in \mathbb{C}$ for which $\{\gamma_n(z, \{\psi_k\})\}_{n=1}^{\infty}$ given by*

$$\gamma_n(z, \{\psi_k\}) := \frac{|\lambda_n| + Re(\lambda_n e^{-i(\psi_n + \psi_{n-1})})}{2Re[(z - c_n)e^{-i\psi_n}] \cdot Re[(z - c_{n-1})e^{-i\psi_{n-1}}]} \tag{3.6}$$

is a chain sequence for some real sequence $\{\psi_n\}$.

Corollary 3.6. *Let λ_n and c_n be real for all n, and let M_1, M_2 and N be three positive constants such that*

$$-M_1^2 \leq \lambda_n \leq M_2^2 \quad \text{and} \quad -N \leq c_n \leq N \quad \text{for all } n.$$

Then all zeros of all $P_n(z)$ are contained in the domain D consisting of all $z = x + iy \in \mathbb{C}$ such that

$$\begin{cases} 2\max\{M_2(|x| - N), M_1|y|\} > (|x| - N)^2 + y^2 & \text{if } |x| - N \geq 0, \\ |y| < 2M_1 & \text{if } |x| - N < 0. \end{cases} \tag{3.7}$$

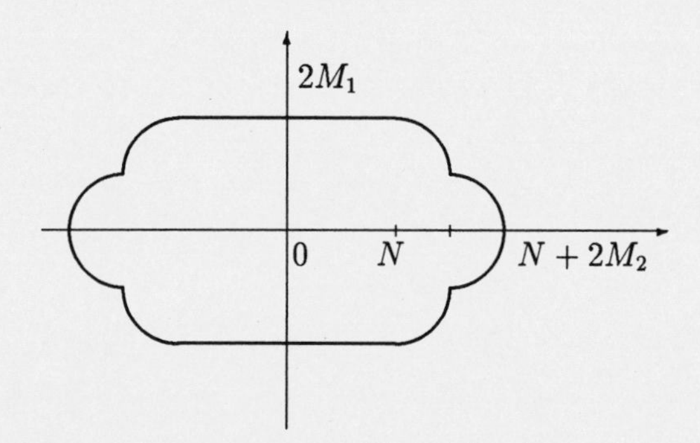

Figure 1

This domain D is symmetric about the real and the imaginary axis, it is bounded and (depending on the constants M_1, M_2 and N) may look something like Fig. 1.

Proof of Corollary 3.6. Let $z = x + iy \in \mathbb{C} \setminus D$ be fixed, and choose

$$\psi_n = \psi = \begin{cases} \arg(z - N) & \text{if } x \geq N, \\ \arg(z + N) & \text{if } x \leq -N, \\ \frac{\pi}{2}\operatorname{sgn}(y) & \text{if } -N < x < N. \end{cases}$$

Assume first that $x \geq N$. Then $\cos\psi > 0$, so that

$$\begin{aligned} \gamma_n(z, \{\psi\}) &= \frac{|\lambda_n| + \lambda_n \cos 2\psi}{2\left[|z - N| + (N - c_n)\cos\psi\right]\left[|z - N| + (N - c_{n-1})\cos\psi\right]} \\ &\leq \frac{|\lambda_n| + \lambda_n \cos 2\psi}{2|z - N|^2}. \end{aligned}$$

Hence all $\gamma_n(z, \{\psi\}) \leq 1/4$, and thus $\{\gamma_n(z, \{\psi\})\}$ is a chain sequence if

$$\frac{M_2^2 \cos^2\psi}{|z - N|^2} \leq \frac{1}{4} \quad \text{and} \quad \frac{M_1^2 \sin^2\psi}{|z - N|^2} \leq \frac{1}{4}.$$

This clearly holds since $e^{i\psi} = \dfrac{z - N}{|z - N|} = \dfrac{x - N + iy}{|z - N|} = \cos\psi + i\sin\psi$.

The proof goes similarly if $x \leq -N$. If $-N < x < N$, then $\cos\psi = 0$ so that

$$\gamma_n(z, \{\psi\}) = \frac{|\lambda_n| - \lambda_n}{2y^2} \leq \frac{1}{4} \quad \text{if} \quad \frac{M_1^2}{y^2} \leq \frac{1}{4}.$$

∎

Let us finally look at the cartesian ovals of Example 2.3. If we choose all $\Gamma_n = 0$ in (2.13), then we are back to the Worpitzky case of Example 2.1. If we choose all $\Gamma_n = \Gamma$ and all $\rho_n = |\Gamma + 1/2|$ in (2.13), and let $\Gamma \to \infty$ in such a way that $\arg(\Gamma(1 + \Gamma)) = 2\psi$ is kept constant, then we are back to a version of the parabola case in Example 2.2. (See [4, Theorems 8.1C, 8.2C, p. 120-122].) So, the oval theorem is closely connected to these two previous cases.

Observation 3.7. *Let all $c_n = 0$ and let all λ_n be contained in the cartesian oval*

$$E := \{\lambda\zeta \in \mathbb{C};\ |\zeta - 1| + s|\zeta| \leq t\} \tag{3.8}$$

for given numbers $\lambda \in \mathbb{C}$, $0 < s < 1$ and $t \geq 1$. Then all zeros of all $P_n(z)$ are contained in the domain

$$D := \left\{z \in \mathbb{C};\ 1 \leq \left|\frac{1 + \sqrt{1 - 4\lambda/(z^2(1 - s^2))}}{1 - \sqrt{1 - 4\lambda/(z^2(1 - s^2))}}\right| < \frac{t}{s}\right\}. \tag{3.9}$$

Proof. Let $z \in \mathbb{C} \setminus D$. Then

$$\Gamma := \frac{z}{2}(\sqrt{1-4\lambda/(z^2(1-s^2))}-1), \text{ where } \mathrm{Re}\sqrt{\ldots} \geq 0$$

satisfies

$$\left|\frac{z+\Gamma}{\Gamma}\right| \geq \frac{t}{s} > 1 \quad \text{and} \quad \Gamma(z+\Gamma)(1-s^2) = -\lambda.$$

For $\lambda_n = \lambda\zeta_n \in E$ it therefore follows that

$$-\lambda_n = \Gamma(z+\Gamma)(1-\frac{\rho^2}{|z+\Gamma|^2})\zeta_n \quad \text{where} \quad \rho = s|z+\Gamma|$$

and

$$|\zeta_n - 1| + s|\zeta_n| = |\zeta_n - 1| + |\zeta_n|\frac{\rho}{|z+\Gamma|} \leq t \leq \frac{\rho}{|\Gamma|}$$

so that

$$|\lambda_n(\overline{z+\Gamma}) - \lambda(\overline{z+\Gamma})| + |\lambda_n|\rho \leq \rho(|z+\Gamma|^2 - \rho^2)$$

and the result follows from (2.11)-(2.12). ■

Remarks 3.8. 1. The method still works if $0 < s < t < 1$, but then $0 \notin V_n$, where V_n is given by (2.11). This means that we get zero-free regions for $P_n^*(z, w(z))$ where $|w(z) - \Gamma(z)| \leq \rho(z) = s|z + \Gamma(z)|$.

2. If we allow $c_n \neq 0$ we can still apply the same idea. Clearly, $(-\lambda_n, z - c_n)$ satisfies (2.12) with $\Gamma = \Gamma(z - c_n)$, $\rho = \rho(z - c_n)$ as above, if $z - c_n \in \mathbb{C}\backslash D$ and $\lambda_n \in E$. Hence, if $\lambda_n \in E$ and all $c_n \in G$ for some complex set G, then no $P_n(z)$ has any zeros in

$$H = \bigcap_{c \in G} \{z;\ z - c \in \mathbb{C}\backslash D\}.$$

4. Tail sequences and chain sequences

Let $\{\alpha_n\}$ be the chain sequence given by (1.13) with parameter sequence $\{g_n\}$. That is, $\alpha_n = (1-g_{n-1})g_n$ for all n. Further, let $t_n := -(1-g_n)$ for $n = 0,1,2,\dots$. Then $-\alpha_n = t_{n-1}(1+t_n)$ for all n, so that $\{t_n\}$ is a tail sequence for the continued fraction $\mathrm{K}(-\alpha_n/1)$. Hence, $\{\alpha_n\}$ is a chain sequence if and only if $\mathrm{K}(-\alpha_n/1)$ has a tail sequence $\{t_n\}$ with

$$-1 \le t_0 < 0 \quad \text{and} \quad -1 < t_n < 0 \quad \text{for } n = 1,2,3,\dots . \tag{4.1}$$

Several of the known properties for parameter sequences of chain sequences are really properties of tail sequences for continued fractions. Let me mention some examples:

Theorem 4.1. (Auric 1907 [1], [8, Satz 2.45, p. 96], Waadeland 1984 [11], J. 1986 [3]). *Let $\{t_n\}$ be a tail sequence for the continued fraction $\mathrm{K}(a_n/b_n)$ with all $t_n \neq \infty$. Then $\mathrm{K}(a_n/b_n)$ converges if and only if*

$$\sum_{k=0}^{\infty}\prod_{n=1}^{k}\frac{b_n+t_n}{-t_n} \tag{4.2}$$

converges to a value $P \in \hat{\mathbb{C}}$. If $\mathrm{K}(a_n/b_n)$ converges, then it converges to $t_0(1-1/P)$. For every $v_0 \in \hat{\mathbb{C}}$,

$$v_N := t_N(1+1/T_N); \quad T_N := \frac{t_0}{v_0-t_0}\prod_{n=1}^{N}\frac{-t_n}{b_n+t_n} + \sum_{k=0}^{N-1}\prod_{n=k+1}^{N}\frac{-t_n}{b_n+t_n} \tag{4.3}$$

for N=0,1,2,...

is also a tail sequence for $\mathrm{K}(a_n/b_n)$.

Proof. From formula (2.3) in [3] it follows that

$$\frac{1}{S_N(w)-t_0} = \frac{1}{t_0}\left[\frac{t_N}{w-t_N}\prod_{n=1}^{N}\frac{b_n+t_n}{-t_n} - \sum_{k=0}^{N-1}\prod_{n=1}^{k}\frac{b_n+t_n}{-t_n}\right].$$

Setting $w := 0$ proves the first part, and $w := v_N$, so that $S_N(v_N) = v_0$, proves the last part. ∎

Theorem 4.2. (Pincherle 1894 [9], [6, Theorem 5.7, p. 164], J. 1986 [3].) *Let $\{t_n\}$ and $\{u_n\}$ be two tail sequences for the continued fraction $\mathrm{K}(a_n/b_n)$ with $t_0 \neq u_0$ and all $t_n, u_n \neq \infty$. Then $\mathrm{K}(a_n/b_n)$ converges if and only if*

$$\prod_{n=0}^{\infty} t_n/u_n \tag{4.4}$$

converges to a value $Q \in \hat{\mathbb{C}}$.

If $\mathrm{K}(a_n/b_n)$ converges, then it converges to $(t_0-u_0Q)/(1-Q)$. For every $v_0 \in \hat{\mathbb{C}}$,

$$v_N := \frac{u_N(v_0-t_0)-t_NQ_N(v_0-u_0)}{v_0-t_0-Q_N(v_0-u_0)}; \quad Q_N := \prod_{n=0}^{N-1}\frac{t_n}{u_n} \tag{4.5}$$

is also a tail sequence for $\mathrm{K}(a_n/b_n)$.

Proof. From formula (2.4) in [3] it follows that

$$\frac{S_N(w)-t_0}{S_N(w)-u_0}=\frac{w-t_N}{w-u_N}Q_N.$$

Hence, the result follows by choosing $w:=0$ and $w:=v_N$. ∎

Remark 4.3. In the proof of Observation 3.7 we used that $t_n=\Gamma$ for all n gives a tail sequence for the continued fraction $\mathrm{K}((-\lambda/(1-s^2))/z)$. (Confer with the fact that $g_n=1/2$ for all n is a parameter sequence for the chain sequence $\alpha_n=1/4$ for all n.)

References

[1] A. Auric, *Recherches sur les fractions continues algébraiques*, J. Math. Pures Appl. (6) **3** (1907).

[2] T. S. Chihara, *Introduction to Orthogonal Polynomials*, Mathematics and Its Applications Ser., Gordon, 1978.

[3] L. Jacobsen, *Composition of linear fractional transformations in terms of tail sequences*, Proc. Amer. Math. Soc. (1) **97** (1986), 97-104.

[4] L. Jacobsen and W. J. Thron, *Oval convergence regions and circular limit regions for continued fractions* $\mathrm{K}(a_n/1)$, Lecture Notes in Math., Springer-Verlag **1199** (1986), 90-126.

[5] W.B. Jones, W. J. Thron, *Convergence of continued fractions*, Canad. J. Math. **20** (1968), 1037-1055.

[6] W.B. Jones, W.J. Thron, *Continued Fractions. Analytic Theory and Applications.* Encyclopedia of mathematics and its applications, **11**, Addison-Wesley, 1980.

[7] R.E. Lane, *The convergence and values of periodic continued fractions*, Bull. Amer. Math. Soc. **51** (1945), 246-250.

[8] O. Perron, *Die Lehre von den Kettenbrüchen*, Band II, Teubner, Stuttgart, 1957.

[9] S. Pincherle, *Delle funzioni ipergeometriche e di vari questioni ad esse attinenti*, Giorn. Mat. Battaglini **32** (1894), 209-291.

[10] W.T. Scott, H.S. Wall, *A convergence theorem for continued fractions*, Trans. Amer. Math. Soc **47** (1940), 155-172.

[11] H. Waadeland, *Tales about tails*, Proc. Amer. Soc. (1) **90** (1984), 57-64.

[12] H.S. Wall, *Analytic Theory of Continued Fractions*, Van Nostrand, New York, 1948.

[13] J.D. Worpitzky, *Untersuchungen über die Entwicklung der monodromen und monogenen Funktionen durch Kettenbrüche*, Friedrichs-Gymnasium und Realschule, Berlin, Jahresbericht 1865, 3-39.

Received: August 7, 1989

Computational Methods and Function Theory
Proceedings, Valparaíso 1989
St. Ruscheweyh, E.B. Saff, L. C. Salinas, R.S. Varga (*eds.*)
Lecture Notes in Mathematics **1435**, pp. 103–115

On Thurston's Formulation and Proof of Andreev's Theorem

Al Marden[1]

University of Minnesota, Minneapolis
Minnesota 55455, USA

and

Burt Rodin[2]

University of California, San Diego
La Jolla, California 92093, USA

In Chapter 13 of his notes [4], W. Thurston states a general result, Theorem 13.7.1 (see also Corollary 13.6.2), regarding the existence and uniqueness of circle packings of prescribed combinatorial type on closed surfaces. This theorem treats the cases of genus $g = 1$ and $g \geq 2$; it is pointed out that the case $g = 0$, which is not proved in these notes, is a result of E.M. Andreev [1,2].

It is implicit in Thurston's notes that the continuity method used there to prove Theorem 13.7.1 in the cases of genus $g = 1$ and $g \geq 2$ could be modified to give a proof of the $g = 0$ case. Such a proof would be very different form Andreev's. The purpose of the present paper is to present such a proof. We separate the statement of the $g = 0$ case into two parts: Theorem A below, which deals with standard circle packings, and Theorem B, which allows the circles to intersect at prescribed angles. These theorems have applications to conformal mapping [3].

1. A *circle packing* on the Riemann sphere or in the plane is a collection of closed disks or the Riemann sphere on in the Euclidean plane with the property that the interiors of the disks are disjoint. The *nerve* of such a circle packing is the graph which has a vertex for each disk and an edge connects two vertices if and only if the corresponding closed disks intersect.

Thurston's theorem, in the case of circle packings, is the following:

[1]Research partially supported by the NSF.
[2]Research partially supported by the NSF and DARPA/ACMP.

Theorem A. *Let $\mathcal{T}$ be a triangulation of the Riemann sphere P. There exists a circle packing of P whose nerve is isomorphic to the one dimensional skeleton of $\mathcal{T}$; any two such circle packings are images of each other under some linear fractional transformation or its complex conjugate.*

2. We begin the proof with some formulas for the Euler characteristic. Let V, E, and F denote the number of vertices, edges, and faces in the triangulation $\mathcal{T}$. Then

$$V - E + F = 2. \tag{1}$$

Since $3F = 2E$, we can eliminate E in (1) and obtain

$$2V = F + 4. \tag{2}$$

It will be convenient to label the vertices of the triangulation $\mathcal{T}$ by $v_1, v_2, \ldots, v_V$.

Let $r = (r_1, r_2, \ldots, r_V)$ be a vector of V positive numbers. Then r determines a polygonal structure on the topological 2-sphere $|\mathcal{T}|$ as follows. Associate to each face of $\mathcal{T}$, with vertices (v_i, v_j, v_k) say, the Euclidean triangle determined by the centers of three mutually (externally) tangent circles of radii r_i, r_j and r_k. Transfer the Euclidean metric on this Euclidean triangle to the associated face of $\mathcal{T}$. Note that the metric is well defined on an edge which is common to two different faces. In this way $|\mathcal{T}|$ becomes a locally Euclidean space with cone type singularities at the vertices; we denote this space by $\mathcal{T}_r$.

The *curvature of $\mathcal{T}_r$ at the vertex v_i*, denoted by $\kappa_r(v_i)$, is defined as follows. Consider all faces of $\mathcal{T}_r$ which have v_i as one of their vertices. Let $\sigma(v_i)$ be the sum of each angle at v_i in each of these triangles. Then

$$\kappa_r(v_i) \equiv 2\pi - \sigma(v_i). \tag{3}$$

Let us note that

$$\sum_{i=1}^{V} \kappa_r(v_i) = 4\pi. \tag{4}$$

Indeed, the left hand side reduces to $2\pi V - \sum \theta$, where $\sum \theta$ denotes the sum of all angles of all triangles of $\mathcal{T}_r$. This sum is equal to πF. An application of Equation (2) now establishes (4).

3. If $\lambda > 0$ then $\mathcal{T}_r$ and $\mathcal{T}_{\lambda r}$ are similar in the sense that corresponding angles are equal. Therefore $\kappa_r(v_i) = \kappa_{\lambda r}(v_i)$. It turns out to be advantageous to normalize the map

$$r \to f(r) = (\kappa_r(v_1), \kappa_r(v_2), \ldots, \kappa_r(v_V))$$

by restricting its domain to the simplex

$$\Delta = \{(r_1, r_2, \ldots, r_V)) \in R^V : r_1 > 0, r_2 > 0, \ldots r_V > 0 \ \&\ r_1 + r_2 + \ldots + r_V = 1\}. \tag{5}$$

It follows form (4) that the range of f can be taken as the hyperplane

$$Y = \{(y_1, y_2, \ldots, y_V) \in R^V : y_1 + y_2 + \ldots + y_V = 4\pi\}. \tag{6}$$

4. For convenience of notation, assume that v_1, v_2, v_3 are the vertices of a single face τ_0 of $\mathcal{T}$. We now prove that the existence assertion of Theorem A will follow once it is shown that the point

$$p_0 = (4\pi/3, 4\pi/3, 4\pi/3, 0, \dots, 0),$$

for example, lies in the image of the map

$$f : \Delta \to Y.$$

To see this, suppose $f(r_0) = p_0$. If we remove that face τ_0 from $\mathcal{T}_{r_0}$ then the remaining triangles can be placed isometrically in the plane, one by one, in an orientation preserving manner, keeping identified edges coincident. Since the curvature is zero at each interior vertex of this complex it can be shown that we obtain in this way an isometric embedding of $\mathcal{T}_{r_0}$ less τ_0 onto a triangle in the plane. {To prove that this is so, one can first show that the process of placing adjacent faces in the plane yields a well defined isometry once the image of an initial face is fixed. For suppose a sequence of adjacent faces are placed in the plane in this way and suppose the first face in the sequence is the same as the last. Then the placement of the first and of the last face will agree–this is clearly true if the sequence of faces surrounds only one interior vertex of $\mathcal{T}_{r_0}$ less τ_0, and can be shown to be true in general by induction on the number of such vertices. The second step in the proof is to use the fact that this placement process provides a locally isometric embedding of $\mathcal{T}_{r_0}$ less τ_0 into the plane and is an actual embedding of the boundary of $\mathcal{T}_{r_0}$ less τ_0. It is easy to see that a local embedding of a topological disk into the plane which is an actual embedding on the boundary must by a global embedding. One concludes that this placement process is a global isometric embedding of $\mathcal{T}_{r_0}$ less τ_0 onto a triangle in the plane}.

We have constructed an isometric embedding, call it ϕ, of $\mathcal{T}_{r_0}$ less τ_0 onto a triangle ABC in the plane. It follows from the definition of $\mathcal{T}_{r_0}$ that if we center a circle of radius r_i at the point $\phi(v_i)$ we obtain a circle packing in the plane whose nerve is isomorphic to the one dimensional skeleton of $\mathcal{T}$. Stereographic projection transforms this packing to a packing of the Riemann sphere with the same property.

It will be useful when we discuss uniqueness to observe that triangle ABC is necessarily equilateral. To verify this, weld another copy of triangle ABC to this one along corresponding edges. One then obtains an isometric image of all of $\mathcal{T}_{r_0}$. We can calculate the curvature at the vertex $\phi(v_1)$ which, we may assume, corresponds to the point A, directly from the definition (3) using this isometric image. We see that the curvture at A is 2π less the sum $\sigma(A)$ of all angles in this isometric image with this vertex A. This sum $\sigma(A)$ is clearly twice the angular measure $m(A)$ of angle A in triangle ABC. On the other hand, we know by the definition of p_0 that the curvature must turn out to be $4\pi/3$. Thus $4\pi/3 = 2\pi - 2m(A)$. Hence $m(A) = \pi/3$. Similarly, $m(B) = m(C) = \pi/3$ and so ABC is equilateral.

5. We now show that $f : \Delta \to Y$ is one to one. Let $r' = (r'_1, r'_2. \dots, r'_V)$ and $r'' = (r''_1, r''_2, \dots, r''_V)$ be distinct points in Δ. Let $\mathcal{V}_0$ be the set of vertices v_i of $\mathcal{T}$ for which $r'_i < r''_i$. Note that the definition (5) of Δ implies that $\mathcal{V}_0$ is a nonempty proper subset of the set of all vertices of $\mathcal{T}$.

Consider a vertex $v \in \mathcal{V}_0$ together with all the faces of $\mathcal{T}_r'$ which have v as vertex. In each such face there is an angle at v, and we classify this angle as type α if it is the only angle in this face which has its vertex in $\mathcal{V}_0$, of type β if two vertices in this face are in $\mathcal{V}_0$, and of type γ if all three vertices of this face are in $\mathcal{V}_0$. Now

$$\sum_{v \in \mathcal{V}_0} \kappa_{r''}(v_i) = \sum_{v \in \mathcal{V}_0} (2\pi - \sigma(v)) \tag{7}$$

$$= 2\pi|\mathcal{V}_0| - \sum(\angle s \text{ of type } \alpha) - \sum(\angle s \text{ of type } \beta) - \sum(\angle s \text{ of type } \gamma).$$

Consider three mutually tangent circles in the plane and their triangle of centers. If one of the circles shrinks and the other two either expand or stay the same size, and if the three circles always remain mutually tangent, then the angle in the triangle of centers with vertex at the center of the shrinking circle will (strictly) increase. If two of the circles shrink and the other either expands or stays the same size, then in the triangle of centers the sum of the two angles which have their vertices at the centers of the shrinking circles will increase. These observations show that if r'' is replaced by r' in equations (7) then

$$\sum_{v \in \mathcal{V}_0} \kappa_{r''}(v_i) > \sum_{v \in \mathcal{V}_0} \kappa_{r'}(v_i). \tag{8}$$

Indeed, in passing from r'' to r' the radii at $v \in \mathcal{V}_0$ shrink and so the first two quantities in

$$\sum(\angle s \text{ of type } \alpha),\ \sum(\angle s \text{ of type } \beta),\ \sum(\angle s \text{ of type } \gamma)$$

will each increase, and the third will remain constant. Since $\mathcal{V}_0$ is a nonempty proper subset of vertices, not all angles are of type γ. It follows that the inequality in (8) is strict and that $f : \Delta \to Y$ must be one-to-one.

6. We now examine the behaviour of $f(r)$ as r tends to a boundary point $s = (s_1, s_2, \dots, s_v)$ of Δ. It will turn out–and this seems very remarkable–that f cannot be extended continously to the boundary of Δ, yet the set of accumulation points of $f(r)$ as r tends to the boundary of Δ form the boundary of a polyhedron. Let $\mathcal{V}_0$ be the set of vertices v_i in $\mathcal{T}$ for which $s_i = 0$; $\mathcal{V}_0$ is a nonempty proper subset of $\mathcal{V}$. We classify the angles of $\mathcal{T}_r$ into types α, β, γ as above. Then as $r \to s$ we have

$$\begin{aligned} \sum(\angle s \text{ of type } \alpha) &\to \pi|\alpha|, \\ \sum(\angle s \text{ of type } \beta) &\to \pi|\beta|/2, \\ \sum(\angle s \text{ of type } \gamma) &\to \pi|\gamma|/3, \end{aligned} \tag{9}$$

where $|x|$ denotes the number of angles of type x. Therefore equation (7) yields

$$\lim_{r \to s} \sum_{v \in \mathcal{V}_0} \kappa_r(v) = 2\pi|\mathcal{V}_0| - \pi|\alpha| - \frac{\pi|\beta|}{2} - \frac{\pi|\gamma|}{3} \tag{10}$$

$$= 2\pi|\mathcal{V}_0| - \pi \cdot (\text{no. of faces with a vertex in } \mathcal{V}_0).$$

From (8) and (10) we see that the image $f(\Delta)$ of $f : \Delta \to T$ lies in the bounded convex polyhedron Y_0 formed by intersecting Y with the half spaces

$$\sum_{i\in I} y_i > 2\pi|I| - \pi \cdot (\text{no. of faces with a vertex in } \mathcal{V}_I \equiv \{v_i : i \in I\}) \tag{11}$$

as I varies over all nonempty proper subsets of $\{1, 2, \dots, V\}$. We have also seen that the accumulation points of $f(r)$ as $r \to \partial\Delta$ lie on the hyperplanes

$$\sum_{i\in I} y_i = 2\pi|I| - \pi \cdot (\text{no. of faces with a vertex in } \mathcal{V}_I \equiv \{v_i : i \in I\}). \tag{12}$$

which form the boundary of Y_0.

7. We know that $f : \Delta \to Y_0$ is a continuous 1-1 mapping. Hence, by Invariance of the Domain, f is a homeomorphism. We also know that $f(r) \to \partial Y_0$ as $r \to \partial\Delta$. It follows by elementary topology that $f : \Delta \to Y_0$ is surjective. Indeed, merely pass to the one point compactifications and apply the simple fact that if X, Y are Hausdorff spaces with X compact and connected and Y connected, and if $\phi : X \to Y$ is continuous and open, then ϕ is surjective.

8. We complete the proof of the existence part of Theorem A by showing that $p_0 = (4\pi/3, 4\pi/3, 4\pi/3, 0, \dots, 0)$ is in the image Y_0 of $f : \Delta \to Y$ (see Section 4). According to (11), this can be done by showing that for every nonempty proper subset I of $\{1, 2, \dots, V\}$,

$$\sum_{i\in I} p_i > 2\pi|I| - \pi \cdot (\text{no. of faces with a vertex in } \mathcal{V}_I \equiv \{v_i : i \in I\}), \tag{13}$$

where $p_0 = (p_1, p_2, \dots, p_V) = (4\pi/3, 4\pi/3, 4\pi/3, 0, \dots, 0)$.

If $|I| = V - 1$ then every face has a vertex in $\mathcal{V}_I \equiv \{v_i : i \in I\}$. Therefore the right hand side of (13) is, by (2),

$$2\pi(V-1) - \pi \cdot F = 2\pi. \tag{14}$$

For subsets I of this cardinality the left hand side of (13) becomes

$$\sum_{i\in I} p_i = 8\pi/3 \text{ or } 12\pi/3. \tag{15}$$

Thus p_0 satisfies (13) when $|I| = V - 1$.

If $|I| = V - 2$ similar reasoning shows that the right hand side of (13) is zero while the left hand side is at least $4\pi/3$. Thus p_0 satisfies (13) in this case also.

We shall show that p_0 satisfies (13) when $1 \le |I| \le V - 3$ by proving that the right hand side of (13) will be negative in these cases. First we rewrite the right hand side in a more invariant form. Let F_1, F_2, F_3 denote, respectively, the number of faces of $\mathcal{T}$ which have exactly 1,2,3 vertices in $\mathcal{V}_I$. Then the right side of (13) is $\pi(2|I| - F_1 - F_2 - F_3)$. Let

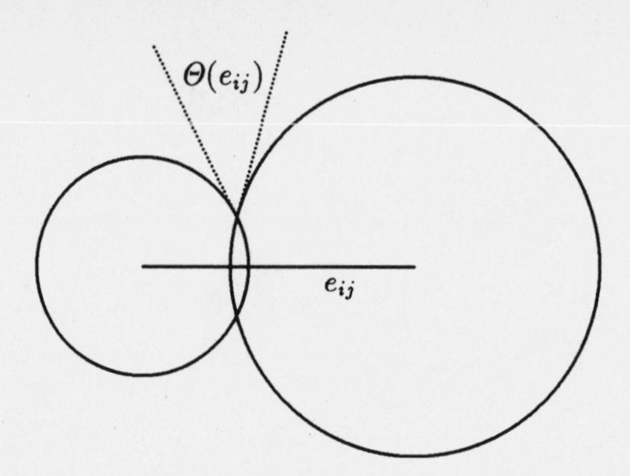

Figure 1. The angle of intersection of two disks

E_2 denote the number of edges of $\mathcal{T}$ which have both of their boundary vertices in $\mathcal{V}_I$. Then the simplicial complex $\mathcal{T}_I$ spanned by the vertices of $\mathcal{V}_I$ has Euler characteristic $\chi_0 = |I| - E_2 + F_3$. Since $3F_3 + F_2 = 2E_2$, we can eliminate E_2 from the expression for χ_0 and obtain $2\chi_0 = 2|I| - F_3 - F_2$. Therefore the condition (13) that $(p_1, p_2, \ldots, p_V)$ lies in the image of f can be rewritten as

$$\sum_{i \in I} p_i > \pi(2\chi_0 - F_1) \tag{17}$$

for every nonempty proper subset I of $\{1, 2, \ldots, V\}$.

We wish to show that the right hand side of (17) is negative for $1 \leq |I| \leq V - 3$. For that purpose we may assume that $\mathcal{T}_I$ is connected. The Euler characteristic of a connected simplicial 2-complex can be interpreted as $2 - 2g - n$ where g is the genus and $n = 1, 2, \ldots$ is the connectivity. In our case $g = 0$ and $\chi_0 = 2 - n$. If $n \geq 3$ there is nothing to prove. If $n = 1$ or 2, one of the components of the complement of $\mathcal{T}_I$ contains at least two vertices in $\mathcal{V} - \mathcal{V}_I$. Therefore we can find an edge (x', y') where the vertices x' and y' are in $\mathcal{V} - \mathcal{V}_I$. We can even choose x' and y' so that there is an edge (y', a') with a' in $\mathcal{V}_I$. If we examine the star of y' we can find a triangle face (x, y, a) of $\mathcal{T}$ with $a \in \mathcal{V}_I$ and $x, y \in \mathcal{V} - \mathcal{V}_I$.

Now look at the union of the star of x and the star of y. If all the vertices adjacent to x and y belong to $\mathcal{V}_I$ the $\{x, y\}$ is a component of the complement of $\mathcal{T}_I$. Since there are at least three vertices in $\mathcal{V} - \mathcal{V}_I$ we must be in the case $n = 2$. Therefore the right hand side of (17) is negative in this case since (x, y, a) is an F_1 type triangle. In the remaining case the set of vertices adjacent to x and to y contains $a \in \mathcal{V}_I$ and some vertex z $(\neq x, y) \in \mathcal{V} - \mathcal{V}_I$. It follows that there are three F_1 type triangles in the union of the stars of x and y.

9. The proof of the existence part of Theorem A is now complete. We have seen that a given triangulation $\mathcal{T}$ of the Riemann sphere P can be realized as the nerve of a circle packing of P. By means of a linear fractional transformation of P we can always arrange the realization so that any three preassigned mutually tangent circles will have

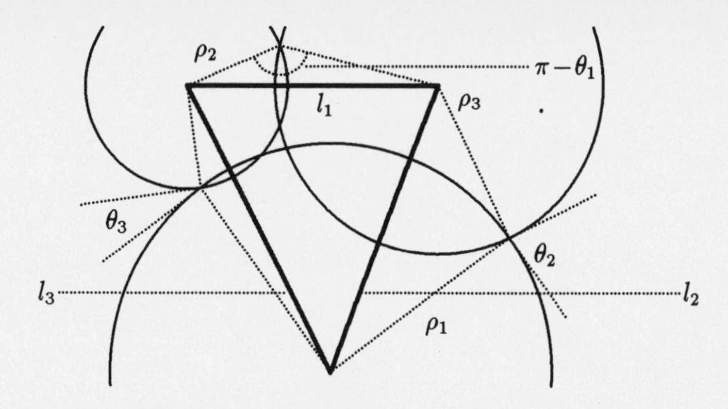

Figure 2. Prescribed intersection angles

equal radii; ∞ will be an interior point of the face whose vertices are their centers. The nerve of this packing on the finite complex plane forms a triangulation of an equilateral triangle by straight line segments.

Let the circles which correspond to the vertices $v_1, v_2, \ldots, v_V$ have radii $r_1, r_2, \ldots, r_V$ respectively. We may assume that $r_1 + r_2 + \ldots + r_V = 1$. For $r_0 = (r_1, r_2, \ldots, r_V)$ we can calculate the curvatures $f(r_0)$ of $\mathcal{T}_{r_0}$ by means of this triangulated equilateral triangle as was done in the last paragraph of Section 4. If v_1, v_2, v_3 correspond to the vertices of the equilateral triangle we find that

$$f(r_0) = (4\pi/3, 4\pi/3, 4\pi/3, 0, \ldots, 0).$$

By the injectivity of f, r_0 is uniquely determined. Therefore the radii of the circles in this normalized circle packing are uniquely determined. It is clear that two circle packings with the same abstract nerve and with corresponding radii equal are (proper of improper) rigid motions of each other. This proves that all circle packings which realize the same triangulation of a 2-sphere are linear fractional transformations of each other followed possibly by a reflection. This completes the proof of Theorem A.

10. We now consider a generalization of Theorem A in the spirit of Theorem 13.7.1 in Thurston's notes *(loc. cit)* for the case $g = 0$. Let $\mathcal{T}$ be a triangulation of the 2-sphere, let $\mathcal{E}$ be the set of edges in $\mathcal{T}$, and let $\Theta : \mathcal{E} \to [0, \pi/2]$ be any function. A family $\mathcal{C}$ of closed disks on the Riemann sphere P or in the plane will be said to *realize* the data $\mathcal{T}, \Theta$ provided the following conditions are satisfied: (a) the nerve of $\mathcal{C}$ is isomorphic to $\mathcal{T}$, and (b) two disks C_i and C_j in $\mathcal{C}$ intersect if and only if their angle of intersection has radian measure $\Theta(e_{ij})$, where e_{ij} is the edge of $\mathcal{T}$ which spans the vertices corresponding to C_i and C_j. (The *angle of intersection* of two disks is the one in the exterior of the two disks; see Figure 1.) We shall prove the following result.

Theorem B. *Let $\mathcal{T}$ be a triangulation of the 2-sphere. Let $\Theta : \mathcal{E} \to \mathbf{R}$ be a function defined on the edges of $\mathcal{T}$ with the property that $0 \le \Theta(e) \le \pi/2$ for all $e \in \mathcal{E}$. Assume that Θ has the following two properties: (i) If $e_1 + e_2 + e_3$ is a cycle of edges in $\mathcal{T}$ then $\Theta(e_1) + \Theta(e_2) + \Theta(e_3) < \pi$, and (ii) if $e_1 + e_2 + e_3 + e_4$ is a cycle of distinct edges in $\mathcal{T}$*

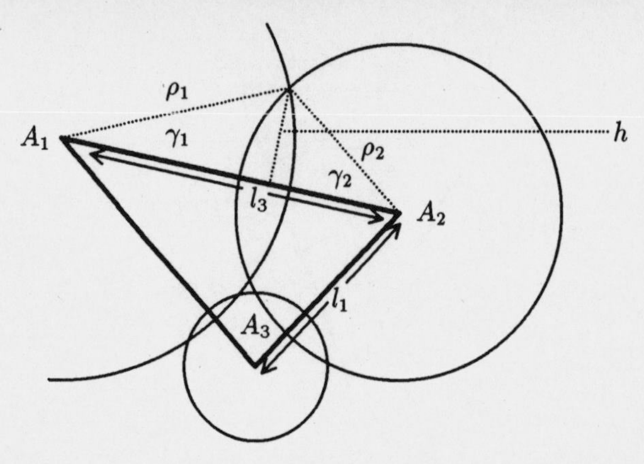

Figure 3

then $\Theta(e_1)+\Theta(e_2)+\Theta(e_3)+\Theta(e_4) < 2\pi$. Then there exists a family $\mathcal{C}$ of round disks on the Riemann sphere which realizes the data $\mathcal{T}, \Theta$. This family $\mathcal{C}$ is uniquely determined up to a linear fractional transformation or its conjugate.

11. The proof proceeds as before except for several additional complications. The first complication arises in constructing $\mathcal{T}_r$. The metric on a face with vertices v_i, v_j, v_k should be the Euclidean metric of the triangle of centers of three disks which intersect pairwise at the nonobtuse angles $\Theta(e_{ij})$, $\Theta(e_{jk})$, and $\Theta(e_{ki})$ and which have the radii prescribed by r. The following lemma from [4] guarantees the existence of such a configuration.

Lemma 2. *For any three nonobtuse angles $\theta_1, \theta_2, \theta_3$ and any three positive numbers ρ_1, ρ_2, ρ_3, there is a unique configuration in the plane consisting of three disks having these radii and intersecting in these angles.*

The proof of Lemma 2 refers to Figure 2. Determine the sides l_1, l_2, l_3 of the desired triangle of centers as follows. Side l_1 is the length of the third side of a triangle which has sides ρ_2 and ρ_3 and included angle $\pi - \theta_1$. Sides l_1 and l_2 are determined similarly. To see that l_1, l_2, l_3 satisfy the triangle inequality note that property (i) of Theorem B implies $l_1 \leq \rho_2+\rho_3 \leq l_2+l_3$, and similarly for l_2 and l_3, because the angles of intersection are nonobtuse. Thus the configuration shown in Figure 2 can always be constructed and is uniquely determined.

12. Having constructed $\mathcal{T}_r$ we proceed as before to define the curvatures at the vertices $v_1, v_2, \ldots, v_V$ and thereby obtain the map

$$r \to f(r) = (\kappa_r(v_1), \kappa_r(v_2), \ldots, \kappa_r(v_V)) : \Delta \to \Sigma. \tag{18}$$

The previous proof (Section 5) that f is one to one can be imitated with the help of the following lemma from [4].

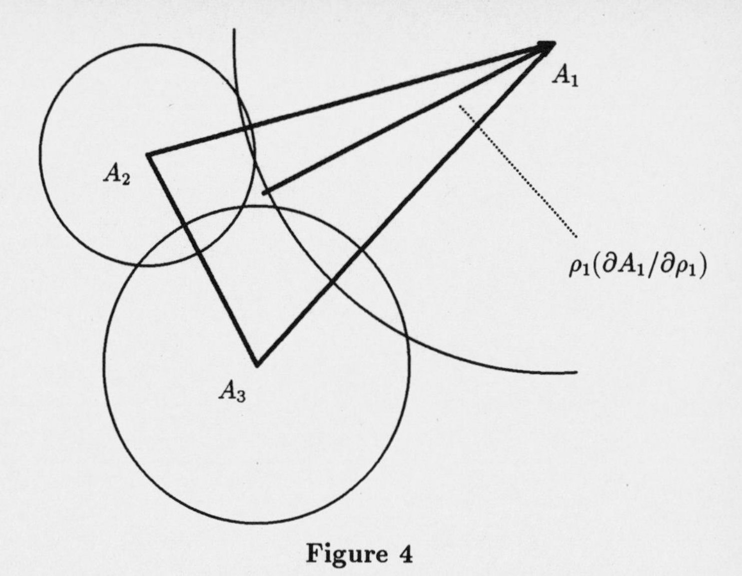

Figure 4

Lemma 3. *Consider three circles in the plane which intersect pairwise in nonobtuse angles. If one radius decreases and the other two remain the same, and if the circles continue to intersect each other at the original angles then, in the triangle of centers, the angle at the vertex of the shrinking circle will increase and the other two angles will decrease.*

Let the radii be ρ_1, ρ_2, ρ_3 and let the triangle of centers have sides of lengths l_1, l_2, l_3 and vertices at A_1, A_2, A_3. If we fix A_2 as the origin, so $\|A_1\| = l_3$, then we can differentiate $A_1 = l_3 u$, where u is a unit vector, to obtain

$$\frac{\partial A_1}{\partial \rho_1} = \frac{\partial l_3}{\partial \rho_1} \frac{A_1}{\|a_1\|} + l_3 B, \tag{19}$$

where B is orthogonal to A_1. Note that (Figure 3) $l_3 = \rho_1 \cos \gamma_1 + \rho_2 \cos \gamma_2$ and so

$$\frac{\partial l_3}{\partial \rho_1} = -\rho_1 \frac{\gamma_1}{\partial \rho_1} \sin \gamma_1 - \rho_2 \frac{\gamma_2}{\partial \rho_1} \sin \gamma_2 + \cos \gamma_1. \tag{20}$$

Since $\rho_2 \sin \gamma_2 = \rho_1 \sin \gamma_1 \equiv h$, (20) becomes

$$\frac{\partial l_3}{\partial \rho_1} = -h \left(\frac{\partial \gamma_1}{\partial \rho_1} + \frac{\partial \gamma_2}{\partial \rho_1} \right) + \cos \gamma_1. \tag{21}$$

The term in parentheses is zero since $\gamma_1 + \gamma_2$ is a constant equal to the fixed angle of intersection of circles 1 and 2. Therefore $\dfrac{\partial l_3}{\partial \rho_1} = \cos \gamma_1$. Thus

$$\rho_1 \frac{\partial l_3}{\partial \rho_1} = \rho_1 \cos \gamma_1, \tag{22}$$

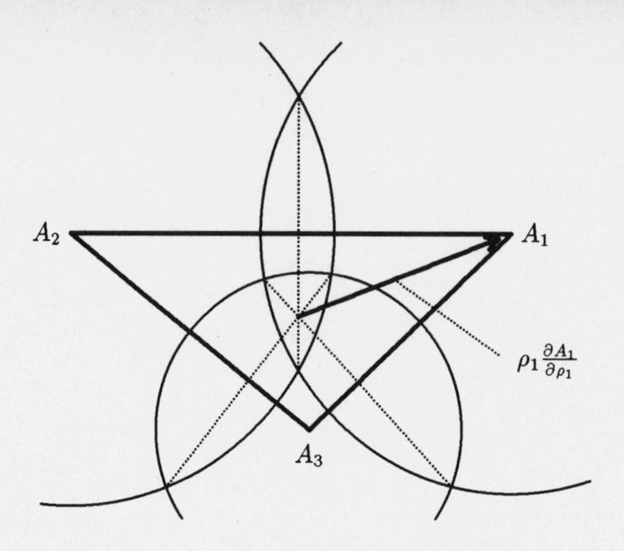

Figure 5

which is the distance form A_1 to the point of intersection of the radical axis of circles 1 and 2 (that is, the line through their points of intersection) with the line joining their centers. It follows that $\rho_1 \dfrac{\partial A_1}{\partial \rho_1}$ is the vector with its tail at A_1 and its tip on the radical axis of circles 1 and 2. By symmetry, its tip is also on the radical axis of circles 1 and 3. Hence $\rho_1 \dfrac{\partial A_1}{\partial \rho_1}$ is the vector from the common point of intersection of the three radical axes to the point A_1 (Figures 4 and 6; in the context of Theorem B the configuration of Figure 6 will not occur because condition (ii) in that theorem insures that the three disks have empty intersection).

Therefore, if ρ_1 decreases and ρ_1 and ρ_2 remain unchanged, the vertex A_1 will move toward the intersection point Q of the radical axes. If we show that Q lies in the triangle of centers it will follow that the angle at A_1 decreases and the proof of Lemma 3 will be complete.

If Q did not lie in the triangle of centers then one of the sides of that triangle would separate Q from the vertex not on that side. Suppose side A_2A_3 separates Q from A_1 as in Figure 6. First note that the circle centered at A_1 does not intersect side A_2A_3 because if it did then the distance A_2A_3 would be greater than the sum of lengths of the tangents from A_2 and A_3 to circle 1 and the lengths of these tangents are upper bounds for r_2 and r_3 since circles 2 and 3 must intersect circle 1 in nonobtuse angles (see Figure 6). Thus in Figure 7 the circle centered at A_1 in the upper half plane determined by A_2A_3 cannot enter the lower plane. It follows that the radical axis of circles 1 and 2 cannot enter the shaded region of Figure 7. This contradicts the fact that Q lies on that radical axis. Therefore Q must lie in the triangle of centers.

13. We now consider the modifications to Section 6 that are needed for the present case. Consider three circles which intersect pairwise in nonobtuse angles. Let the radius

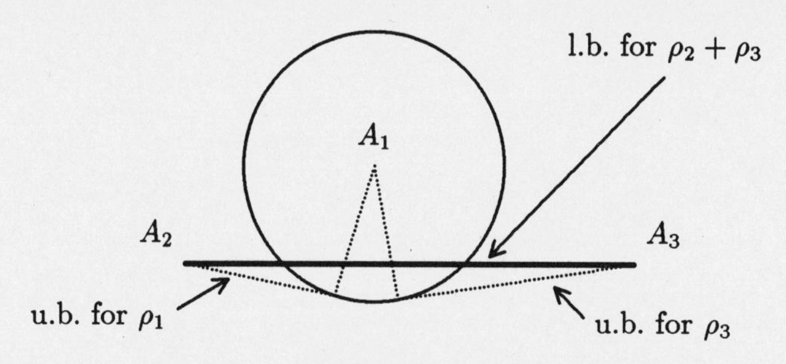

Figure 6

of circle 1 shrink to zero while the intersection angles and the other two radii remain constant. Then in the triangle of centers the angle at the center of circle 1 will increase to the limiting value $\pi - (\theta(e))$, where e is the opposite edge. If the radii of two circles shrink to zero then in the triangle of centers the sum of the two angles at the centers of these two circles will tend to the limiting value π. When all three radii shrink to zero we use the fact that the sum of the three angles is constantly equal to π. Thus the three equations (9) are to be replaced by

$$\begin{aligned} &\textstyle\sum(\angle s \text{ of type } \alpha) \rightarrow \sum(\pi - \Theta(e(\alpha)), \\ (23)\qquad &\textstyle\sum(\angle s \text{ of type } \beta) \rightarrow \pi|\beta|/2, \\ &\textstyle\sum(\angle s \text{ of type } \gamma) \rightarrow \pi|\gamma|/3. \end{aligned}$$

Equation (10) is replaced by

$$(24)\qquad \lim_{r\to s} \sum_{v\in\mathcal{V}_0} \kappa_r(v) = 2\pi|\mathcal{V}_0| - \sum(\pi - \Theta(e(\alpha)) - \frac{\pi|\beta|}{2} - \frac{\pi|\gamma|}{3}$$

$$= 2\pi|\mathcal{V}_0| - \pi \cdot \text{ (no. of faces with a vertex in } \mathcal{V}_0) + \sum_{\alpha} \Theta(e(\alpha)).$$

We conclude that the image Y_0 of $f : \Delta \rightarrow Y$ is the hyperplane formed by intersecting Y with the half spaces

$$(25)\qquad \sum_{i\in I} y_i > 2\pi|I| - \pi \cdot \text{ (no. of faces with a vertex in } \mathcal{V}_I) + \sum_{\alpha} \Theta(e(\alpha))$$

for each nonempty proper subset I of $\{1, 2, \dots, V\}$.

14. As in Section 8, the existence assertion of Theorem B will follow from showing that $p_0 = (4\pi/3, 4\pi/3, 4\pi/3, 0, \dots, 0)$ is in the image Y_0 of $f : \Delta \rightarrow Y$. According to (25) this is equivalent to showing that for each nonempty proper subset I of $\{1, 2, 3, \dots, V\}$,

$$(26)\qquad \sum_{i\in I} p_i > 2\pi|I| - \pi \cdot \text{ (no. of faces with a vertex in } \mathcal{V}_I) + \sum_{\alpha} \Theta(e(\alpha))$$

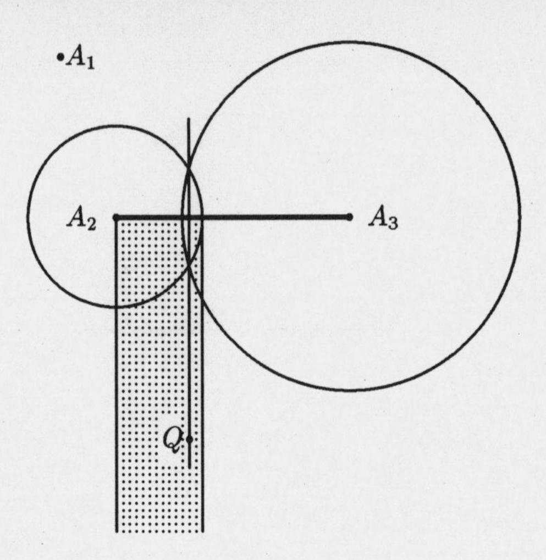

Figure 7

where $p_0 = (p_1, p_2, \ldots, p_V) = (4\pi/3, 4\pi/3, 4\pi/3, 0, \ldots, 0)$ and where the summation denotes the sum over angles of type α of the value of Θ at the edge opposite the angle of type α.

If $|I| = V - 1$, (26) holds. Indeed, there are no angles of type α in this case, so the earlier calculation (Equations (14) and (15)) applies.

If $|I| = V - 2$ the left hand side of (26) is at least $4\pi/3$. The first two terms on the right hand side cancel by (2) and the summation term is at most π since there can be at most two terms in the sum. Thus (26) holds in this case as well.

For the cases $1 \leq |I| \leq V - 3$, rewrite (26) in the form

$$\sum_{i \in I} p_i > \pi(2\chi_0 - F_1) + \sum_{\alpha} \Theta(e(\alpha)) = 2\pi\chi_0 - \sum_{e} (\pi - \Theta(e)) \tag{27}$$

in the same way that (13) was rewritten as (17); here the sum of F_1 terms (each term satisfies $\pi/2 \leq \pi - \Theta(e) \leq \pi$) is taken over the edges in the F_1 type triangles opposite the angles of type α. These edges form a 1-cycle in the mod 2 homology of $\mathcal{T}$.

As before, we have to show that the right hand side of (27) is negative and we may assume that the complex spanned by $\mathcal{V}_I$ is connected ($\chi_0 \leq 1$). The negativity is clear if $\chi_0 = 0$. The reasoning in Section 8 following (17) showed that if $\chi_0 = 0$ then $F_1 \geq 1$, and if $\chi_0 = 1$ then $F_1 \geq 3$. Therefore the right hand side is negative if $\chi_0 = 0$; it will also be negative if $\chi_0 = 1$ and $F_1 \geq 5$. The remaining cases $\chi_0 = 1$ and $F_1 = 3$ or 4 are covered by properties (i) and (ii) of Theorem B. In case $F_1 = 4$ one can eliminate the case that the four edges are not distinct because in that case $\mathcal{V} - \mathcal{V}_I$ will consist of the vertices on these edges. Since they do not span a triangle face, one of the vertices v_1, v_2, v_3 is in $\mathcal{V}_I$. Therefore the left hand side of (27) is at least $4\pi/3$ and the desired inequality (27) holds, even though the right hand side may be zero.

References

[1] E.M. Andreev, *On convex polyhedra in Lobacevskii spaces*, Math. USSR Sb. **10** (1970), 413-440.

[2] E.M. Andreev, *On convex polyhedra of finite volume in Lobacevskii space*, Mat. Sb., Nor. Ser. **83** (1970), 256-260 (Russian); Math. USSR, Sb. **12** (1970), 255-259 (English).

[3] B. Rodin, D. Sullivan, *The convergence of circle packings to the Riemann mapping*, J. Diff. Geom. **26** (1987), 349-360.

[4] W.P. Thurston, *The Geometry and Topology of 3-manifolds*, Princeton University Notes, Princeton, New Jersey, 1980.

[5] W.P. Thurston, *The finite Riemann mapping theorem*, invited address, International Symposium in Celebration of the Proof of the Bieberbach Conjecture. Purdue University, March 1985.

Received: December 27, 1989

Computational Methods and Function Theory
Proceedings, Valparaíso 1989
St. Ruscheweyh, E.B. Saff, L. C. Salinas, R.S. Varga (*eds.*)
Lecture Notes in Mathematics **1435**, pp. 117–129

Hyperbolic Geometry in Spherically k-convex Regions

Diego Mejía

Departamento de Matemáticas, Universidad Nacional
Apartado Aéreo 568, Medellín, Colombia

and

David Minda[1, 2]

Department of Mathematics, University of Cincinnati
Cincinnati, Ohio 45221-0025, USA

1. Introduction

This paper is the second of a proposed trilogy dealing with the concept of k-convexity in various geometries. The first paper [MM] dealt with k-convexity in euclidean geometry. This paper presents a generalization of most of the results of [MM] to k-convexity in spherical geometry. In a proposed third paper we plan to treat the concept of k-convexity in hyperbolic geometry.

We assume that the reader is familiar with [MM] and frequently omit details of proofs when they are similar to proofs of analogous results in [MM]. In fact, since many of the proofs in [MM] are truly geometric in nature, it is often clear how to extend the proof from euclidean geometry to the case of spherical geometry. A Jordan region Ω on the Riemann sphere P is called spherically k-convex if the spherical curvature of the boundary is at least k at each point of $\partial\Omega$. This assumes that the boundary of Ω is smooth. A definition of spherical k-convexity that applies to arbitrary regions is given in section 3. Our program is to give sharp spherical estimates for various hyperbolic quantities in spherically k-convex regions. These estimates lead to sharp distortion and covering theorems (including the Bloch-Landau constant and the Koebe set) for the

[1]Research partially supported by NSF Grant No. DMS-8801439.

[2]I want to thank the University of Cincinnati and the Taft Foundation for partially supporting my sabbatical leave during the 1988-89 academic year and the University of California, San Diego for its hospitality during this period.

family $K_s(k,\alpha)$ of normalized ($f(0)=0$, $f'(0)=\alpha$) conformal mappings of the unit disk $\mathbf{D}$ onto spherically k-convex regions. For the special case of spherically convex regions (the case $k=0$) some of these results were established in [M₁] and [M₃].

2. Spherical geometry

We begin by recalling some basic facts about spherical geometry on the Riemann sphere $\mathbf{P}$; see [M₁] for more details. The spherical metric is $\lambda_{\mathbf{P}}(z)|dz| = |dz|/(1+|z|^2)$; it has Gaussian curvature $+4$. The group of rotations of the Riemann sphere is

$$\operatorname{Rot}(\mathbf{P}) = \left\{ T(z) = \frac{e^{i\theta}(z-a)}{1+\bar{a}z} : \theta \in \mathbf{R},\ a \in \mathbf{P} \right\}.$$

The spherical metric is invariant under the group $\operatorname{Rot}(\mathbf{P})$; that is, $T^*(\lambda_{\mathbf{P}}(z)|dz|) = \lambda_{\mathbf{P}}(z)|dz|$, or, equivalently, $|T'|/(1+|T|^2) = 1/(1+|z|^2)$. The spherical distance between $a,b \in \mathbf{P}$ is defined by

$$d_{\mathbf{P}}(a,b) = \inf \int_\gamma \lambda_{\mathbf{P}}(z)|dz|,$$

where the infimum is taken over all paths γ on $\mathbf{P}$ which join a and b. Moreover, this infimum is actually a minimum and is attained for the shorter arc δ of any great circle through a and b. The arc δ is unique unless a and b are antipodal points in which case both of the subarcs of any great circle through a and b is a possible choice for δ. Any path δ between a and b such that

$$d_{\mathbf{P}}(a,b) = \int_\delta \lambda_{\mathbf{P}}(z)|dz|$$

is called a spherical geodesic. The explicit formula for the spherical distance is

$$d_{\mathbf{P}}(a,b) = \begin{cases} \arctan(|a-b|/|1+\bar{b}a|) & \text{if } a,b \in \mathbf{C} \\ \arctan(1/|a|) & \text{if } a \in \mathbf{C},\ b = \infty. \end{cases}$$

Observe that $d_{\mathbf{P}}(a,b) \le \pi/2$. Also, the spherical distance is invariant under the group of rotations of $\mathbf{P}$. Sometimes it is more convenient to employ a related quantity in place of the spherical distance. Set $E_{\mathbf{P}}(a,b) = \tan d_{\mathbf{P}}(a,b)$ and note that this quantity is also invariant under the group $\operatorname{Rot}(\mathbf{P})$. Let $D_s(a,\varphi)$ denote the spherical disk with center a and radius φ.

We shall also make use of the notion of spherical curvature. We briefly recall this concept; for more details the reader should consult [M₁]. Suppose γ is a C^2 curve on $\mathbf{P}$ with nonvanishing tangent at the point $a \in \mathbf{P}$. The spherical curvature of γ at a is

$$k_s(a,\gamma) = \frac{k(a,\gamma) - (\partial/\partial n)\log \lambda_{\mathbf{P}}(a)}{\lambda_{\mathbf{P}}(a)},$$

where $k(a,\gamma)$ is the euclidean curvature of γ at a and $n = n(a)$ is the unit normal at a which makes an angle $+\pi/2$ with the tangent vector at a. For example, if γ is the spherical circle $\{z \in \mathbf{P} : d_{\mathbf{P}}(a,z) = \varphi\}$, $\varphi \in (0,\pi/2)$, oriented so that the center a lies on the left-hand side of γ, then $k_s(z,\gamma) = 2\cot(2\varphi)$. Observe that the spherical curvature of γ is positive for $\varphi \in (0, \pi/4)$ and negative for $\varphi \in (\pi/4, \pi/2)$; the value $\varphi = \pi/4$ yields a great circle which is a spherical geodesic and has zero spherical curvature. If f is a meromorphic function and γ is a path, then for $z \in \gamma$

$$k_s(f(z), f\circ\gamma)f^{\#}(z) = k_s(z,\gamma)\lambda_{\mathbf{P}}(z) + \operatorname{Im}\left(\left[\frac{2\bar{z}}{1+|z|^2} + \frac{f''(z)}{f'(z)} - \frac{2\overline{f(z)}f'(z)}{1+|f(z)|^2}\right] t(z)\right),$$

where $t(z)$ is the unit tangent to γ at z. Recall that $f^{\#}(z) = |f'(z)|/(1+|f(z)|^2)$ denotes the spherical derivative of a meromorphic function.

Now we turn to an issue involving hyperbolic geometry for regions on the Riemann sphere. For a hyperbolic region Ω on the Riemann sphere $\mathbf{P}$ it is natural to consider the *spherical density* of the hyperbolic metric in place of the density of the hyperbolic metric. For a hyperbolic region Ω on the Riemann sphere the spherical density is

$$\mu_\Omega(z) = \frac{\lambda_\Omega(z)|dz|}{\lambda_{\mathbf{P}}(z)|dz|} = (1+|z|^2)\lambda_\Omega(z)$$

which is a continuous function on Ω and is invariant under rotations of $\mathbf{P}$.

We shall frequently make use of the spherical distance to the boundary of the region. Let $\varepsilon_\Omega(z) = \min\{d_{\mathbf{P}}(z,c) : c \in \partial\Omega\}$. This is the spherical distance from z to $\partial\Omega$ and is clearly invariant under rotations of $\mathbf{P}$. In some applications our formulas become much simpler if we employ a related quantity in place of $\varepsilon_\Omega(z)$. Define $E_\Omega(z) = \tan\varepsilon_\Omega(z)$. This quantity is invariant under rotations of $\mathbf{P}$.

We reformulate several basic results for the hyperbolic metric in terms of the spherical density. These results are well known for the euclidean density of the hyperbolic metric.

Principle of Hyperbolic Metric. *Suppose Ω is a hyperbolic region on $\mathbf{P}$. If f is meromorphic on the unit disk $\mathbf{D}$ and $f(\mathbf{D}) \subset \Omega$, then $\mu_\Omega(f(z))f^{\#}(z) \le \lambda_{\mathbf{D}}(z) = 1/(1-|z|^2)$ for $z \in \mathbf{D}$ with equality if and only if f is a conformal mapping of $\mathbf{D}$ onto Ω.*

Monotonicity Property. *Suppose Ω and Δ are hyperbolic regions on $\mathbf{P}$ and $\Omega \subset \Delta$. Then $\mu_\Delta(z) \le \mu_\Omega(z)$ for $z \in \Omega$ with equality if and only if $\Omega = \Delta$.*

3. Geometric properties of spherically k-convex regions

In this section we introduce the concept of a spherically k-convex region and study some of its basic properties. The proofs of some of the results are so similar to the analogous results for euclidean k-convex regions that we sometimes omit details; see [MM] for the details in the case of euclidean k-convexity.

Suppose that $k > 0$, $a, b \in \mathbf{P}$ and $d_{\mathbf{P}}(a,b) < \arctan(2/k)$. Then there are two distinct closed spherical disks $\overline{D}_1$ and $\overline{D}_2$, each of radius $\frac{1}{2}\arctan(2/k)$, such that $a, b \in \partial\overline{D}_j$ $(j = 1,2)$; note that $\partial\overline{D}_j$ has constant spherical curvature k. Let $S_k[a,b] = \overline{D}_1 \cap \overline{D}_2$. The boundary of $S_k[a,b]$ consists of two closed circular arcs Γ_1 and Γ_2, each with constant spherical curvature k. We define $S_0[a,b]$ to be the spherical geodesic between a and

b when $d_{\mathbf{P}}(a,b) < \pi/2$. Also, for $d_{\mathbf{P}}(a,b) = \arctan(2/k)$ we let $S_k[a,b]$ be the closed spherical disk with center at the midpoint of the spherical geodesic joining a and b and radius $\frac{1}{2}\arctan(2/k)$. Then for $0 \le k' < k \le 2/\tan(d_{\mathbf{P}}(a,b))$,. $S_{k'}[a,b] \subset S_k[a,b]$.

Definition. Let $k \in [0,\infty)$. A region $\Omega \subset \mathbf{P}$ is called spherically k-convex if for any pair of points $a,b \in \Omega$ with $d_{\mathbf{P}}(a,b) < \arctan(2/k)$ we have $S_k[a,b] \subset \Omega$.

Clearly, spherically 0-convex is equivalent to spherical convexity, so we shall employ the phrase *spherical k-convexity* only when $k > 0$ and use *spherical convexity* instead of *spherical 0-convexity*. Note that if Ω is spherically k-convex, then it is also spherically k'-convex for $0 \le k' \le k$. In particular, a spherically k-convex region is always spherically convex and simply connected. For each $k > 0$ any spherical disk of radius $\frac{1}{2}\arctan(2/k)$ is spherically k-convex but not spherically k'-convex for any $k' > k$. The intersection of a finite number of spherically k-convex regions is spherically k-convex and the union of an increasing sequence of spherically k-convex regions is again spherically k-convex.

Lemma 0. *Suppose that Ω is a simply connected region on $\mathbf{P}$ with $d_{\mathbf{P}}(a,b) < \pi/2$ for all $a,b \in \Omega$. If at each point $c \in \partial\Omega$ there is a supporting great circle for all points of Ω in a sufficiently small neighborhood of c, then Ω is spherically convex.*

Proof. This proof is adaptation of the proof of the analogous result for euclidean convexity [S]. Let $a,b \in \Omega$. We want to show that the spherical geodesic connecting a and b is contained in Ω. By rotating Ω if necessary, we may asume that $a = 0$. Because Ω contains no antipodal points, Ω must lie in $\mathbf{C}$. Then 0 and b can be joined by a polygonal curve Π with straight sides which is contained in Ω. Let $0 = z_0, z_1, z_2, \ldots, z_m, z_{m+1} = b$ be the vertices of Π in the order in which they are met in traversing Π from 0 to b. We show that the vertices can be removed one at time, so that eventually the polygon, while remaining inside Ω, becomes the straight line $[0,b]$, the spherical geodesic connecting 0 to b. Suppose that $[0,z_k] \subset \Omega$. We want to show that $[0,z_{k+1}] \subset \Omega$. If 0, z_k and z_{k+1} are collinear, we are done. Suppose $[0,z_{k+1}] \not\subset \Omega$. Consider the set of all segments $[0,p]$, where p ranges over $[z_k, z_{k+1}]$; let $\theta(p)$ denote the angle between $[0,p]$ and $[0,z_k]$. There is a smallest angle $\theta(p_0)$ such that $p_0 \in (z_k, z_{k+1}]$ and $[0,p_0]$ contains a point of $\partial\Omega$. Let $c \in \partial\Omega \cap [0,p_0]$ be the point in this set that is closest to the origin. Then all points of the closed euclidean triangle Δ with vertices $0, z_k$ and p_0 are in Ω, except for c and possibly other points of $[c,p_0]$. But then the point c fails to have a supporting great circle locally since the great circle through 0 and c contains points of Ω arbitrarily close to c and any other great circle through c meets the triangle Δ inside. ∎

Proposition 1. *Suppose that Ω is a simply connected region on $\mathbf{P}$ bounded by a closed C^2 Jordan curve $\partial\Omega$ such that $d_{\mathbf{P}}(a,b) < \pi/2$ for all $a,b \in \Omega$. If $k_s(c,\partial\Omega) \ge k > 0$ for all $c \in \partial\Omega$, then Ω is spherically k-convex.*

Proof. We begin by showing that Ω is spherically convex. By the preceding lemma, it is sufficient to show that there is a locally supporting great circle at each point $c \in \partial\Omega$. We may assume that $c = 0$. Then $k(0,\partial\Omega) = k_s(0,\partial\Omega) > 0$. Hence, $k(\zeta,\partial\Omega) > 0$ for all $\zeta \in \partial\Omega$ in a neighbourhood of 0, so Ω has a locally supporting euclidean straight line at 0 [S, p. 46]. This straight line through 0 is also a great circle, so Ω has a locally supporting great circle at c.

Next, we show that Ω is spherically k-convex. Fix $a, b \in \Omega$. Let τ be the supremum of all $t \geq 0$ such that $S_t[a,b] \subset \Omega$. Note that $\tau \leq 2/\tan d_{\mathbf{P}}(a,b)$. Since Ω is spherically convex, we know that $\tau > 0$. We want to show that $\tau > k$. If $\tau = 2/\tan d_{\mathbf{P}}(a,b)$, then Ω contains the closed spherical disk with center at the midpoint of the spherical geodesic joining a and b and radius $\frac{1}{2}d_{\mathbf{P}}(a,b)$. Let D be the largest spherical disk with the same center that is contained in Ω. Then ∂D meets $\partial\Omega$ at some point c. By rotating W if necessary, we may assume $c = 0$. The comparison principle for euclidean curvature [G, p. 28] implies that $k(0,\partial\Omega) \leq k(0,\partial D)$. Consequently, $k \leq k_s(0,\partial\Omega) \leq k_s(0,\partial D) < \tau$. The remaining case is $\tau < 2/\tan d_{\mathbf{P}}(a,b)$. Consider the two circular arcs Γ_1 and Γ_2 of spherical curvature τ which bound $S_\tau[a,b]$. At least one of these two arcs, say Γ_1, meets $\partial\Omega$ in some point c. As before we may assume $c = 0$. The comparison principle for euclidean curvature now gives $k(0,\partial\Omega) \leq k(0,\Gamma_1)$, so that $k \leq \tau$. We need to show strict inequality. Because Ω is open, we can select points a' and b' in Ω so that a and b lie strictly between a' and b' on the spherical geodesic in Ω joining these latter two points. Let τ' be defined relative to a' and b' in the same manner that τ was defined for a and b. Then for a' and b' near a and b, respectively, $\tau' < \tau$. Since $k \leq \tau'$ just as $k \leq \tau$, we obtain $k < \tau$. ■

Proposition 2. *Suppose Ω is a spherically k-convex region. Then for any $a \in \Omega$ and $c \in \partial\Omega$, $S_k[a,c]\backslash\{c\} \subset \Omega$.*

Corollary. *If Ω is a spherically k-convex region, then $\mathrm{int}S_k[c,d] \subset \Omega$ for $c, d \in \partial\Omega$.*

Lemma 1. *Suppose D is an open spherical disk of radius $\frac{1}{2}\arctan(2/k)$ and B is an open spherical disk such that $c \in \partial B \cap \partial D$ and B and D are externally tangent at c. If $d_{\mathbf{P}}(a,c) < \arctan(2/k)$ and $a \notin \overline{D}$, then $S_k[a,b]\backslash\{c\} \cap B \neq \emptyset$.*

Proposition 3. *Suppose Ω is a spherically k-convex region. Assume $a \in \Omega$, $c \in \partial\Omega$ and $d_{\mathbf{P}}(a,c) = \varepsilon_\Omega(a)$. If D is the spherical disk of radius $\frac{1}{2}\arctan(2/k)$ that is tangent to the circle $\{z \in \mathbf{P} : d_{\mathbf{P}}(z,a) = \varepsilon_\Omega(a)\}$ at c and contains a in its interior, then $\Omega \subset D$.*

Proposition 4. *Suppose Ω is a spherically k-convex region. Let $a \in \mathbf{P}\backslash\Omega$, $c \in \partial\Omega$ and $d_{\mathbf{P}}(a,c) = \varepsilon_\Omega(a)$. If D is the spherical disk of radius $\frac{1}{2}\arctan(2/k)$ that is tangent to the circle $\{z \in \mathbf{P} : d_{\mathbf{P}}(z,a) = \varepsilon_\Omega(a)\}$ at c and that does not meet the disk $D_s(a, \varepsilon_\Omega(a))$, then $\Omega \subset D$.*

4. Lower bound for the spherical density of the hyperbolic metric in a spherically k-convex region

We obtain a sharp lower bound for the spherical density in terms of the spherical distance to the boundary of the region.

Theorem 1. *Suppose Ω is a spherically k-convex region. Then for $z \in \Omega$*

$$\mu_\Omega(z) \geq \frac{1 + E_\Omega^2(z)}{E_\Omega(z)[2 - kE_\Omega(z)]}$$

with equality at a point if and only if Ω is a spherical disk with radius $\frac{1}{2}\arctan(2/k)$.

Proof. If D is a spherical disk of radius $\frac{1}{2}\arctan(2/k) = \arctan(\sqrt{k^2+4}-k)/2$, we show that equality holds at each point of D. Since $\mu_\Omega(z)$ is invariant under rotations of $\mathbf{P}$, we may assume without loss of generality that D is centered at the origin. We set $r = (\sqrt{k^2+4}-k)/2$ or $k = (1-r^2)/r$. Then $E_D(z) = (r-|z|)/(1+r|z|)$, or $|z| = (r - E_D(z))/(1+rE_D(z))$, and so

$$\mu_D(z) = (1+|z|^2)\frac{r}{r^2-|z|^2} = \frac{r(1+E_D^2(z))}{E_D(z)[2r-(1-r^2)E_D(z)]} = \frac{(1+E_D^2(z))}{E_D(z)[2-kE_D(z)]}.$$

This shows that equality holds at each point of D.

Now consider any spherically k-convex region Ω. Fix $a \in \Omega$. Select $c \in \partial\Omega$ with $E_\Omega(a) = \tan d_{\mathbf{P}}(a,c)$. Let Γ be the spherical circle with radius $\frac{1}{2}\arctan(2/k)$ through c that is tangent to the spherical disk $D_s(a, \arctan E_\Omega(a))$ at c and whose interior D contains a. Then Proposition 3 implies that $\Omega \subset D$; also $E_D(a) = E_\Omega(a)$. The monotonicity property of the hyperbolic metric yields $\mu_\Omega(a) \geq \mu_D(a)$ with equality if and only if $\Omega = D$. Because $E_D(a) = E_\Omega(a)$, this inequality in conjunction with the above formula for $\mu_D(z)$ completes the proof. ∎

Corollary 1. *Suppose that Ω is a spherically k-convex region. If f is meromorphic in $\mathbf{D}$ and $f(\mathbf{D}) \subset \Omega$, then for $z \in \mathbf{D}$*

$$(1-|z|^2)f^\#(z) \leq \frac{E_\Omega(f(z))[2-kE_\Omega(f(z))]}{1+E_\Omega^2(f(z))}.$$

Equality holds at a point if and only if Ω is a spherical disk with radius $\frac{1}{2}\arctan(2/k)$ and f is a conformal mapping of $\mathbf{D}$ onto Ω.

Proof. The principle of hyperbolic metric gives

$$\mu_\Omega(f(z))f^\#(z) \leq \lambda_{\mathbf{D}}(z) = 1/(1-|z|^2)$$

for $z \in \mathbf{D}$ with equality if and only if f is a conformal mapping of $\mathbf{D}$ onto Ω. The theorem implies that

$$\frac{1+E_\Omega^2(f(z))}{E_\Omega(f(z))[2-kE_\Omega(f(z))]} \leq \mu_\Omega(f(z))$$

with equality if and only if Ω is a spherical disk of radius $\frac{1}{2}\arctan(2/k)$. By combining the two preceding inequalities and the necessary and sufficent conditions for equality, we obtain the corollary. ∎

Definition. Let $K_s(k,\alpha)$ denote the family of all holomorphic functions f defined on $\mathbf{D}$ such that f is univalent, $f(0)=0$, $f'(0)=\alpha$ and $f(\mathbf{D})$ is a spherically k-convex region.

If $f \in K_s(k,\alpha)$ and $\Omega = f(\mathbf{D})$, then the preceding corollary with $z=0$ produces $\alpha = |f'(0)| \leq h(E_\Omega(0))$, where $h(t) = t(2-kt)/(1+t^2)$. Note that $h(t)$ is increasing on the interval $0 \leq t \leq (\sqrt{k^2+4}-k)/2 = r$ and $h(r) = r$. Because Ω is spherically k-convex, we know that $E_\Omega(0) \leq r$. Therefore, $\alpha \leq r$ whenever $f \in K_s(k,\alpha)$. Moreover, $\alpha = r$ if and only if $f(z) = rz$.

Example. Set $f_k(z) = \alpha z/(1 - \sqrt{1 - \alpha(\alpha + k)}\, z)$. Then $f_k \in K_s(k, \alpha)$ since $f_k(\mathbf{D})$ is a spherical disk of radius $\frac{1}{2}\arctan(2/k)$. Note that $f_k(-1) = -\alpha/(1 + \sqrt{1 - \alpha(\alpha + k)})$ and $f_k(1) = \alpha/(1 - \sqrt{1 - \alpha(\alpha + k)})$. The largest spherical disk contained in $f_k(\mathbf{D})$ and centered at the origin has radius $\alpha/(1 + \sqrt{1 - \alpha(\alpha + k)})$.

Corollary 2. *Suppose $f \in K_s(k, \alpha)$. Then either $\{w : |w| \le \alpha/(1+\sqrt{1 - \alpha(\alpha + k)})\}$ is contained in $f(\mathbf{D})$ or $f(z) = e^{-i\theta} f_k(e^{i\theta} z)$ for some $\theta \in \mathbf{R}$.*

Proof. Set $\Omega = f(\mathbf{D})$ and apply the preceding corollary with $z = 0$ to obtain

$$\alpha = |f'(0)| \le \frac{E_\Omega(0)(2 - kE_\Omega(0))}{1 + E_\Omega^2(0)}.$$

This yields $E_\Omega(0) \ge \alpha/(1 + \sqrt{1 - \alpha(\alpha + k)})$ with equality if and only if Ω is a spherical disk of radius $\frac{1}{2}\arctan(2/k)$. In the case of equality, f is a conformal mapping of $\mathbf{D}$ onto a spherical disk of radius $\frac{1}{2}\arctan(2/k)$ that contains the origin and whose boundary is externally tangent to the circle $\{w : |w| = \alpha/(1 + \sqrt{1 - \alpha(\alpha + k)})\}$. In this case it is straightforward to check that f must have the prescribed form. ∎

5. The spherical Bloch-Landau constant for the family $K_s(k, \alpha)$

We derive a sharp lower bound for the spherical density of the hyperbolic metric for a spherically k-convex region in terms of a uniform upper bound on ε_Ω. We use an extremal region which has been employed in [M1] and [MM]. Suppose $k > 0$, $\theta \in (0, \pi/2)$ and $N = \tan\theta$. Let $R = \sqrt{N(2 - kN)/(k + 2N)}$; R is selected so that the circle through $-R, iN$ and R has spherical radius $\frac{1}{2}\arctan(2/k) = \arctan(\sqrt{k^2 + 4} - k)/2$, or equivalently, spherical curvature k. Let $S = S(N) = \mathrm{int} S_k[-R, R]$. Note that for $N = (\sqrt{k^2 + 4} - k)/2$ the set S is actually a spherical disk. In all cases, S contains the disk $\{z : |z| < N\}$, but no larger disk centered at the origin, and S is contained in the disk $D = \{z : |z| < R\}$. Each of the two circular arcs bounding S makes an angle 2φ with the segment $[-R, R]$, where $\varphi = \arctan(N/R)$.

We also introduce a certain collection of *triangular* spherically k-convex regions. Let $\mathcal{T} = \mathcal{T}(N)$ denote the family of all spherically k-convex regions that contain $\{z : |z| < N\}$ and are bounded by three distinct circular arcs each of spherical radius $\frac{1}{2}\arctan(2/k)$ and having the property that the full circles are tangent to $|z| = N$ and contain $\{z : |z| < N\}$ in their interior. Each of these circular arcs will meet ∂D in diametrically opposite points and has euclidean radius $k' = (1 + N^2)/(k + 2N)$. Therefore, each region Δ in $\mathcal{T}$ is both spherically k-convex and euclidean k'-convex. From [MM, Lemma 2] we obtain the following result.

Lemma 2. *If $\Delta \in \mathcal{T}$, then for $z \in \Delta$, $\mu_\Delta(z) > (\pi/4\varphi)\mu_D(z) \ge (\pi/4\varphi R)$.*

Theorem 2. *Suppose Ω is a spherically k-convex region. Let $\theta = \max\{\varepsilon_\Omega(z) : z \in \Omega\}$ and $N = \tan\theta$. Then*

$$\mu_\Omega(z) \geq \frac{\pi}{4}\sqrt{\frac{k+2N}{N(2-kN)}}\frac{1}{\arctan\sqrt{\dfrac{N(k+2N)}{2-kN}}}.$$

Equality holds at a point $a \in \Omega$ if and only if there is a rotation T of $\mathbf{P}$ such that $\Omega = T(S)$ and $a = T(0)$.

Proof. Select $a \in \Omega$ with $E_\Omega(a) = N$. From Proposition 3 we see that

$$\varepsilon_\Omega(a) \leq \frac{1}{2}\arctan\frac{2}{k} = \arctan\frac{\sqrt{k^2+4}-k}{2}$$

with equality if and only if Ω is a spherical disk with center a and radius $\frac{1}{2}\arctan(2/k)$. Hence, $N \leq (\sqrt{k^2+4}-k)/2$ with equality if and only if Ω is a spherical disk with center a and radius $\frac{1}{2}\arctan(2/k)$.

First, suppose $N = (\sqrt{k^2+4}-k)/2$. Then Ω is a spherical disk centered at a and so (see the first part of the proof of Theorem 1)

$$\mu_\Omega(z) = \frac{1+E_\Omega^2(z)}{E_\Omega(z)[2-kE_\Omega(z)]}.$$

The right-hand side of this identity is a strictly decreasing function of $E_\Omega(z)$, so we obtain

$$\mu_\Omega(z) \geq \frac{1+N^2}{N[2-kN]}$$

with strict inequality unless $z = a$. This is the desired result in this case.

Now, assume that $0 < N < (\sqrt{k^2+4}-k)/2$. We may suppose that $a = 0$ since all quantities involved are invariant under rotations of $\mathbf{P}$. Let $I = \{z : |z| = N$ and $z \in \partial\Omega\}$. The set I is nonempty and closed. A result of Blaschke [B] for euclidean convexity readily extends to spherical convexity and implies that I cannot be contained in a closed subarc of the circle $|z| = N$ with angular length strictly less than π. Now the proof completely parallels that of [MM, Thm. 4], so all further details are omitted.■

The function

$$g_k(t) = \sqrt{\frac{t(2-kt)}{k+2t}}\arctan\sqrt{\frac{t(k+2t)}{2-kt}}$$

is strictly increasing on $[0,(\sqrt{k^2+4}-k)/2]$ with maximum value

$$g_k((\sqrt{k^2+4}-k)/2) = \frac{\pi}{4}\left((\sqrt{k^2+4}-k)/2\right).$$

Hence, for $\alpha \in [0,(\sqrt{k^2+4}-k)/2]$ the equation $g_k(t) = \alpha\pi/4$ has a unique solution $N(\alpha) \in [0,(\sqrt{k^2+4}-k)/2]$.

Corollary (Bloch-Landau constant for $K_s(k,\alpha)$). *Let $f \in K_s(k,\alpha)$. Then either $f(\mathbf{D})$ contains an open spherical disk with radius strictly larger than $\arctan N(\alpha)$ or else $f(z) = e^{-i\psi}F(e^{i\psi}z)$ for some $\psi \in \mathbf{R}$, where*

$$F(z) = \sqrt{\frac{N(\alpha)(2-kN(\alpha))}{k+2N(\alpha)}}\tanh\left(\frac{2}{\alpha}\sqrt{\frac{N(\alpha)(2-kN(\alpha))}{k+2N(\alpha)}}\log\frac{1+z}{1-z}\right)$$

belongs to $K_s(k,\alpha)$ *and maps* $\mathbf{D}$ *conformally onto* $S(N(\alpha))$.

Proof. Set $\Omega = f(\mathbf{D})$ and $N = \max\{E_\Omega(z) : z \in \Omega\}$. If $N > N(\alpha)$, then we are done. Assume $N \le N(\alpha)$. Then $g_k(N) \le g_k(N(\alpha)) = \alpha\pi/4$. Since $\lambda_\Omega(0) = 1/f'(0) = 1/\alpha$, the theorem with $z = f(0) = 0$ gives $1/\alpha \ge \pi/4g_k(N)$, or $g_k(N) \ge \alpha\pi/4$. Now $g_k(N) = \alpha\pi/4$, so $N = N(\alpha)$. Thus, equality holds in the theorem at the origin, so Ω is just a rotation of $S(N(\alpha))$. Since $F \in K_S(k,\alpha)$ and maps $\mathbf{D}$ onto $S(N(\alpha))$, we conclude $f(z) = e^{-i\psi}F(e^{i\psi}z)$ for some $\psi \in \mathbf{R}$. ∎

6. Hyperbolic convexity in spherically k-convex regions

We show that the intersection of a spherical disk and a spherically k-convex region is hyperbolically convex provided the center of the disk is sufficiently close to the region.

Example. Let $D = \{z : |z-(i/k)| \le 1/k\}$ and $\Gamma' = \partial D$. Note that D is spherically k-convex since Γ' has constant spherical curvature k. For $a > 0$ we have $E = E_D(-ia) = a$. Suppose G is the spherical circle with center $-ia$ and radius $\arctan\Theta$, where

$$E = \frac{k\Theta^2}{1+\Theta^2+\sqrt{(1+\Theta^2)^2+k^2\Theta^2}}. \tag{1}$$

The euclidean center of Γ is $b = -iE(1+\Theta^2)/(1-\Theta^2E^2)$ and the euclidean radius is $r = \Theta(1+E^2)/(1-\Theta^2E^2)$. Necessary and sufficient for the circles Γ and Γ' to be orthogonal is $r^2 = |b|^2 + (2|b|/k)$. Straightforward calculations show that this condition is equivalent to (1). Therefore, Γ and Γ' are orthogonal, so $\Gamma \cap D$ is a hyperbolic geodesic in D and the hyperbolic half-plane $D \cap \{z : E_{\mathbf{P}}(a,z) < \Theta\}$ is a hyperbolically convex subset of D. On the other hand, if E strictly exceeds the right side of (1), then $D \cap \{z : E_{\mathbf{P}}(a,z) < \Theta\}$ is not a hyperbolically convex subset of D. This example shows that the following theorem is sharp.

Theorem 3. *Suppose* Ω *is a spherically k-convex region on* $\mathbf{P}$. *Let* $a \in \mathbf{P}\backslash\overline{\Omega}$, $E = E_\Omega(a)$, $\Theta > 0$, $R = \{z : E_{\mathbf{P}}(a,z) < \Theta\}$, $\Gamma = \partial R$ *and* j *denote reflection in* Γ. *If*

$$E \le \frac{k\Theta^2}{1+\Theta^2+\sqrt{(1+\Theta^2)^2+k^2\Theta^2}}, \tag{2}$$

then $j(\Omega\backslash R) \subset \Omega$. *In particular,* $\Omega \cap \{z : E_{\mathbf{P}}(a,z) < \Theta\}$ *is a hyperbolically convex subset of* Ω.

Proof. Select $c \in \partial\Omega$ with $E_{\mathbf{P}}(a,c) = E_\Omega(a) = E$. By making use of rotational invariance, we may assume that $c = 0$ and that $\{z : E_{\mathbf{P}}(a,z) < E\}$ is tangent to the real axis at the origin and contained in the lower half-plane. Let D denote the disk of the preceding example and $\Gamma' = \partial D$. Proposition 4 implies $\Omega \subset D$. With the normalization $c = 0$, the proof of Theorem 3 when equality holds in (2) parallels the proof of Theorem 5 of [MM]. Observe that the euclidean and spherical curvature coincide at the origin. As in [MM] the case of strict inequality can be reduced to the case of equality. ∎

7. Applications to spherical curvature

Theorem 4. *Suppose that Ω is a spherically k-convex region, γ is a hyperbolic geodesic in Ω and $z_0 \in \gamma$. Let a denote one of the spherical centers of the spherical circle of curvature for γ at z_0 and Θ the tangent for the spherical radius of this spherical circle of curvature. Then*

$$E_\Omega(a) \geq \frac{k\Theta^2}{1+\Theta^2+\sqrt{(1+\Theta^2)^2+k^2\Theta^2}}.$$

Equality holds if and only if Ω is a spherical disk of radius $\frac{1}{2}\arctan(2/k)$ and γ is a circular arc orthogonal to $\partial\Omega$.

Proof. For $k=0$, the case of spherical convexity, this result was established in [M$_1$]. Since the proof for $k>0$ parallels the case $k=0$ and is similar to the proof of the analogous result for euclidean k-convexity [MM, Thm. 6], the details are omitted. ∎

Corollary 1. *Suppose Ω is a spherically k-convex region, γ is a hyperbolic geodesic in Ω and $z_0 \in \gamma$. Then*

$$|k_s(z_0,\gamma)| \leq \frac{2(1-E_\Omega^2(z_0)-kE_\Omega(z_0))}{E_\Omega(z_0)(2-kE_\Omega(z_0))}.$$

Equality implies that Ω is a spherical disk of radius $\frac{1}{2}\arctan 2k$ and γ is a circular arc orthogonal to $\partial\Omega$.

Proof. There is no loss of generality in assuming that $k_s(z_0,\gamma)>0$. Let Γ be the circle of curvature for γ at z_0. Suppose a is a spherical center of Γ such that the spherical radius of Γ with respect to a is $\varphi \in (0,\pi/4]$. Let $\delta=\varepsilon_\Omega(a)$. The spherical geodesic through z_0 and a meets $\partial\Omega$ at some point c, so we have $\varepsilon_\Omega(z_0) \leq d_{\mathbf{P}}(z_0,c) = d_{\mathbf{P}}(z_0,a)-d_{\mathbf{P}}(c,a) \leq \varphi-\delta$. If $E=E_\Omega(z_0)$, $D=\tan\delta$ and $F=\tan\varphi$, then this gives $D \leq (F-E)/(1+EF)$. The theorem gives

$$\frac{k\Theta^2}{1+\Theta^2+\sqrt{(1+\Theta^2)^2+k^2\Theta^2}} \leq D \leq \frac{F-E}{1+EF}.$$

This yields

$$F \geq \frac{\sqrt{(1+E^2-kE)^2+k^2E^4}-(1-E^2-kE)}{E(2-kE)},$$

so that

$$k_s(z_0,\gamma)=\frac{1-F^2}{F} \leq \frac{2(1-E^2-kE)}{E(2-kE)}.$$

If equality holds here, then equality holds in the preceding inequalities, so Ω is a spherical disk of radius $\frac{1}{2}\arctan(2/k)$ and γ is a circular arc orthogonal to $\partial\Omega$. ∎

Corollary 2. *Let Ω be a spherically k-convex region. Then for $z \in \Omega$*

$$(i) \qquad (1+|z|^2)|\nabla \log \mu_\Omega(z)| \leq \frac{2(1-E_\Omega^2(z)-kE_\Omega(z))}{E_\Omega(z)(2-kE_\Omega(z))}$$

and

$$(ii) \qquad (1+|z|^2)|\nabla \log \mu_\Omega(z)| \le 2\sqrt{[\mu_\Omega(z)]^2 - [1+k\mu_\Omega(z)]}$$

with equality if and only if Ω *is a spherical disk of radius* $\frac{1}{2}\arctan(2/k)$.

Proof. It is straightforward to show that equality holds in both (i) and (ii) for a spherical disk of radius $\frac{1}{2}\arctan(2/k)$. Because of the rotational invariance of the quantities involved, it suffices to establish (i) in the special case $z = 0$. In this case the proof is similar to that of Corollary 2(i) to Theorem 6 in [MM]. We establish (ii) as follows. Theorem 1 gives

$$E_\Omega(z) \ge \frac{1}{\mu_\Omega(z) + \sqrt{[\mu_\Omega(z)]^2 - [1+k\mu_\Omega(z)]}}$$

with equality if and only if Ω is a spherical disk of radius $\frac{1}{2}\arctan(2/k)$. Since the function $h(t) = 2(1-t^2-kt)/[t(2-kt)]$ is strictly decreasing on its domain, we obtain

$$h(E_\Omega(z)) \le h\left(\frac{1}{\mu_\Omega(z) + \sqrt{[\mu_\Omega(z)]^2 - [1+k\mu_\Omega(z)]}}\right) = 2\sqrt{[\mu_\Omega(z)]^2 - [1+k\mu_\Omega(z)]}.$$

Thus, (i) implies (ii). ∎

8. Applications to spherically k-convex mappings

We now apply our results to the study of conformal mappings of $\mathbb{D}$ onto spherically k-convex regions. K.W. Bauer in his thesis [Ba] investigated, among other topics, spherically k-convex mappings and found some of the results given in this section, especially those given in Theorem 6.

Theorem 5. *Suppose* Ω *is a spherically* k*-convex region and* $f : \mathbb{D} \to \Omega$ *is a conformal mapping. Then*

$$\frac{|f''(0)|}{|f'(0)|^2} \le \frac{2[1 - E_\Omega^2(f(0)) - kE_\Omega(f(0))]}{E_\Omega(f(0))[2-kE_\Omega(f(0))]} \le 2\sqrt{[E_\Omega(f(0))]^2 - [1+kE_\Omega(f(0))]}$$

and equality holds if and only if Ω *is a spherical disk of radius* $\frac{1}{2}\arctan(2/k)$.

Proof. Since $|\nabla \log \mu_\Omega(f(0))| = |f''(0)|/|f'(0)|^2$, the result follows from Corollary 2 to Theorem 4. ∎

Corollary. *If* $f \in K_s(k,\alpha)$, *then* $|f''(0)| \le 2\alpha\sqrt{1-\alpha(\alpha+k)}$. *Equality holds if and only if* $f(z) = e^{-i\theta} f_k(e^{i\theta} z)$ *for some* $\theta \in \mathbb{R}$.

Proof. Set $\Omega = f(\mathbb{D})$. Since $f'(0) = \alpha$ and $\mu_\Omega(f(0)) = 1/|f'(0)|$, the result follows from the theorem. Note that the only functions in $K_s(k,\alpha)$ which map onto a spherical disk of radius $\frac{1}{2}\arctan(2/k)$ are the functions $e^{-i\theta} f_k(e^{i\theta} z)$ for some $\theta \in \mathbb{R}$. ∎

Theorem 6. *If $f \in K_s(k,\alpha)$, then for $z \in \mathbb{D}$*

$$\left|\frac{f''(z)}{f'(z)} - \frac{2\bar{z}}{1-|z|^2} - \frac{2\overline{f(z)}f'(z)}{1+|f(z)|^2}\right| \le \frac{2}{1-|z|^2}\sqrt{1-(1-|z|^2)f^{\#}(z)[(1-|z|^2)f^{\#}(z)+k]}.$$

Equality holds at a point if and only if $f(z) = e^{-i\theta} f_k(e^{i\theta} z)$ for some $\theta \in \mathbb{R}$.

Proof. For $z = 0$ this reduces to the preceding corollary. In order to establish the result at the point $a \in \mathbb{D}$, just replace f by

$$g(z) = \frac{f\left(\dfrac{z+a}{1+\bar{a}z}\right) - f(a)}{1+\overline{f(a)}f\left(\dfrac{z+a}{1+\bar{a}z}\right)}.$$

This function is univalent in $\mathbb{D}$, $g(0) = 0$, $g'(0) = (1-|a|^2)f'(a)/(1+|f(a)|^2)$ and $g(\mathbb{D})$ is spherically k-convex. If we apply the preceding corollary to this function g with α replaced by $g'(0)$, then we obtain the desired result. If $f(z) = e^{-i\theta} f_k(e^{i\theta} z)$, then equality holds at $z = re^{-i\theta}$, $0 \le r < 1$. ■

Corollary 1. *Let f be holomorphic and univalent in $\mathbb{D}$ and normalized by $f(0) = 0$ and $f'(0) = \alpha > 0$. Then $f \in K_s(k,\alpha)$ if and only if*

$$1 + \operatorname{Re}\left(\frac{zf''(z)}{f'(z)} - \frac{2z\overline{f(z)}f'(z)}{1+|f(z)|^2}\right) \ge kf^{\#}(z)|z|$$

for $z \in \mathbb{D}$.

Proof. First, suppose $f \in K_s(k,\alpha)$. Straightforward, but tedious, calculations show that the inequality of Theorem 6 implies the inequality of the corollary. On the other hand, suppose that the above inequality holds. Consider the path $\gamma : z = z(t) = re^{it}$, $t \in [0, 2\pi]$. Since $k_s(z,\gamma) = (1-|z|^2)/|z|$, we obtain

$$k_s(f(z), f\circ\gamma) = \frac{1 + \operatorname{Re}\left(\dfrac{zf''(z)}{f'(z)} - \dfrac{2z\overline{f(z)}f'(z)}{1+|f(z)|^2}\right)}{|zf^{\#}(z)|}.$$

Therefore, $k_s(f(z), f\circ\gamma) \ge k$ for all $z \in \gamma$, so $f(\{z : |z| < r\})$ is a spherically k-convex region by Proposition 1. Then $f(\mathbb{D})$ is also spherically k-convex since it is an increasing union of spherically k-convex regions. ■

Corollary 2. *Suppose f is holomorphic and univalent in $\mathbb{D}$. Then $f(\mathbb{D})$ is spherically k-convex if and only if f maps each subdisk of $\mathbb{D}$ onto a spherically k-convex region.*

References

[Ba] K.W. Bauer, *Über die Abschätzung von Lösungen gewisser partieller Differentialgleichungen vom elliptischen Typus*, Bonner Mathematische Schriften, **10**, Bonn, 1960.

[B] W. Blaschke, *Über den größten Kreis in einer konvexen Punktmenge*, Jahresber. Deutsch. Math. Verein. **23** (1914), 369-374.

[G] H.W. Guggenheimer, *Differential Geometry*, McGraw-Hill, New York, 1963.

[MM] D. Mejía, D. Minda, *Hyperbolic geometry in k-convex regions*, Pacific J. Math. (to appear).

[M_1] D. Minda, *The hyperbolic metric and Bloch constants for spherically convex regions*, Complex Variables Theory Appl. **5** (1986), 127-140.

[M_2] D. Minda, *A reflection principle for the hyperbolic metric and applications to geometric function theory*, Complex Variables Theory Appl. **8** (1987), 129-144.

[M_3] D. Minda, *Applications of hyperbolic convexity to euclidean and spherical convexity*, J. Analyse Math. **49** (1987), 90-105.

[S] J.J. Stoker, *Differential Geometry*, Wiley-Interscience, 1969.

Received: May 14, 1989

Computational Methods and Function Theory
Proceedings, Valparaíso 1989
St. Ruscheweyh, E.B. Saff, L. C. Salinas, R.S. Varga (*eds.*)
Lecture Notes in Mathematics **1435**, pp. 131–142

The Bloch and Marden Constants

David Minda[1, 2]

Department of Mathematical Sciences, University of Cincinnati
Cincinnati, Ohio 45221-0025

1. Introduction

Suppose F is holomorphic in the open unit disk $\mathbb{D}$. For $a \in \mathbb{D}$ let $r(a, F)$ denote the largest nonnegative number r such that there is a simply connected region Ω, with $a \in \Omega \subset \mathbb{D}$, which is mapped conformally onto $D(F(a), r)$ by F. (Here we use the notation $D(b, r)$ for the open euclidean disk with center b and radius r. For $b = 0$ we simplify this to $D(r)$.) Set $r(F) = \sup\{r(a, F) : a \in \mathbb{D}\}$. Clearly, $r(a, F) = 0$ if and only if $F'(a) = 0$. The Bloch constant B is defined by

$$B = \inf\{r(F) : F \text{ is holomorphic in } \mathbb{D} \text{ and } F'(0) = 1\}.$$

In 1924 Bloch proved that B is positive in spite of the vast collection of functions over which the infimum is taken. Landau ([L$_1$],[L$_2$]) gave various lower bounds for the Bloch constant the best of which is $B > 0.396$. Landau also showed that

$$B = \inf\{r(F) : F \in \mathcal{B},\ \|F\| = 1 \text{ and } F'(0) = 1\},$$

where $\mathcal{B}$ is the class of all holomorphic functions F defined on $\mathbb{D}$ such that $F(0) = 0$ and

$$\|F\| = \sup\{(1 - |w|^2)|F'(w)| : w \in \mathbb{D}\} < \infty.$$

The normalization $F(0) = 0$ is not really essential but it is convenient. Landau also gave the upper bound $B < 0.555$. Ahlfors and Grunsky [AG] presented a geometric treatment of this example of Landau and noted that Landau did not get the best possible upper bound from his example. They improved the upper bound to

[1]Research partially supported by NSF Grant No. DMS-8801439.

[2]I want to thank the University of Cincinnati and the Taft Foundation for partially supporting my sabbatical leave during the 1988-89 academic year and the University of California, San Diego for its hospitality during this period.

$$B \le .4719\cdots = \frac{\Gamma(1/3)\Gamma(11/12)}{\sqrt{1+\sqrt{3}}\,\Gamma(1/4)}.$$

In addition, they conjectured that this upper bound is the correct value of the Bloch constant. Ahlfors [A_1] developed a powerful differential-geometric method, now called Ahlfors' Lemma and the method of ultrahyperbolic metrics, and, as one application of this method, he showed $B \ge \sqrt{3}/4$. Later, Heins [H_1] introduced the notion of an SK-metric as a generalization of an ultrahyperbolic metric and developed a theory of these metrics, noting the parallel between the theories of subharmonic functions and SK-metrics. In particular, he obtained a sharp form of Ahlfors' Lemma; an application is the improvement of Ahlfors' lower bound to $B > \sqrt{3}/4$. Pommerenke [P], essentially using a function-theoretic version of Ahlfors' Lemma, gave another proof of Heins' lower bound for the Bloch constant and also presented a lower bound for the locally schlicht Bloch constant. Minda [M_1] employed Ahlfors' differential-geometric method to study various Bloch constants. Peschl ([Pe_1], [Pe_2]) obtained a number of results about Bloch constants for families of locally schlicht functions by a different method.

Thus, in the fifty years since Ahlfors' original paper, there has been little progress on improving the lower bound for the Bloch constant. Recently, Bonk [B] introduced a new technique in the study of the Bloch constant. His method yields a sharp lower distortion estimate on $\mathrm{Re}F'$ when $F \in \mathcal{B}$, $\|F\| = 1$ and $F'(0) = 1$. This inequality immediately gives $B \ge \sqrt{3}/4$; intriguingly, the same lower bound that Ahlfors obtained by a different method. By making a more careful use of this distortion theorem, Bonk obtains $B > \sqrt{3}/4 + 10^{-14}$, the first quantitative improvement on the lower bound for the Bloch constant in a half century.

There is another, related constant defined for Bloch functions. This constant is defined relative to the domain of the function while the Bloch constant is defined relative to the range. For $a \in \mathbb{D}$ let $s = s(a,F) \in [0,+\infty]$ be the largest non-negative number such that F is univalent in the hyperbolic disk $D_h(a,s)$. Recall that $d_h(a,w) = 2\,\mathrm{artanh}\,(|w-a|/|1-\bar{a}w|)$ is the hyperbolic distance between the points $a, w \in \mathbb{D}$ and the hyperbolic disk with center a and hyperbolic radius s is $D_h(a,s) = \{w \in D : d_h(a,w) < s\}$. Note that $s(a,F) = 0$ if and only if $F'(a) = 0$ and $s(a,F) = +\infty$ precisely when F is univalent on $\mathbb{D}$. Thus, $s(a,F)$ is the hyperbolic radius of the largest hyperbolic disk centered at a in which F is univalent. Set $s(F) = \sup\{s(a,F) : a \in \mathbb{D}\}$. The Marden constant is defined by

$$M = \inf\{s(F) : 0 < \|F\| < \infty\} = \inf\{s(F) : \|F\| = 1\}.$$

The latter equality holds since $s(F/m) = s(F)$ for any positive constant m. The Marden constant was introduced in [M_3] and named for its analogy with various Marden constants for Fuchsian groups. (The reader is warned that the pseudohyperbolic, rather than the hyperbolic, distance was employed in [M_3].) Minda [M_3] gave the bounds

$$2\,\mathrm{artanh}\,\frac{1}{2} = 1.098\cdots \le M \le 2\,\mathrm{artanh}\,\frac{1}{\sqrt{1+\sqrt{3}}} = 1.401\cdots .$$

The upper bound is conjectured to be sharp and is obtained from the same function that Ahlfors and Grunsky conjecture is extremal for the Bloch constant. The distortion theorem of Bonk immediately gives the improved lower bound $M \geq 2\,\mathrm{artanh}\,1/\sqrt{3} = 1.316\cdots$. Thus, from a single distortion theorem Bonk obtains good lower bounds for both the Bloch and the Marden constants.

The purpose of this paper is to present a different, geometric proof of Bonk's distortion theorem. This new proof yields additional information and seems likely to extend to other situations. In particular, all of the extremal functions for Bonk's distortion theorem are obtained from this new proof; the extremal functions are two-sheeted branched coverings of $\mathbb{D}$ onto another disk. Also, I will give a geometric significance for the constants $\sqrt{3}/4$ and $2\,\mathrm{artanh}\,1/\sqrt{3}$; they give the sharp solution to two geometric problems related to the Bloch and Marden constants. Finally, the extremal functions for Bonk's distortion theorem are related to the ultrahyperbolic metric that is employed in the proof that $B > \sqrt{3}/4$ via Ahlfors' method.

2. Main results

In this section we present a new, geometric proof of Bonk's distortion theorem. The following facts will be used several times. Suppose $F(w)$ is holomorphic in $\mathbb{D}$, $w = S(z)$ is a conformal automorphism of $\mathbb{D}$ and $G = F \circ S$, then $\|F\| = \|G\|$. In fact, the identity

$$|S'(z)| = \frac{1 - |S(z)|^2}{1 - |z|^2}$$

yields the stronger pointwise result

$$(1 - |z|^2)|G'(z)| = (1 - |w|^2)|F'(w)|.$$

Also, $r(S(a), F) = r(a, G)$ and $s(S(a), F) = s(a, G)$ for all $a \in \mathbb{D}$.

We begin by discussing the functions that are extremal for Bonk's distortion theorem.

Example 1. Set

$$f_1(w) = \frac{1 - \sqrt{3}w}{\left(1 - \dfrac{w}{\sqrt{3}}\right)^3} = 1 - \sum_{n=2}^{\infty} \frac{(n^2 - 1)w^n}{(\sqrt{3})^n}$$

and

$$F_1(w) = \int_0^w f_1(\omega)\,d\omega.$$

Note that $f_1(0) = 1$ and $F_1(0) = 0$, $F_1'(0) = 1$. We wish to determine the explicit values of the quantities $\|F_1\|$, $r(0, F_1)$ and $s(0, F_1)$. In order to do this, it is advantageous to express both f_1 and F_1 in another fashion.

Set

$$z = T(w) = \frac{w - \dfrac{1}{\sqrt{3}}}{1 - \dfrac{w}{\sqrt{3}}} \quad \text{and} \quad w = S(z) = T^{-1}(z) = \frac{z + \dfrac{1}{\sqrt{3}}}{1 + \dfrac{z}{\sqrt{3}}}.$$

Note that T and S are conformal automorphisms of $\mathbb{D}$ and

$$T'(w) = \frac{2}{3\left(1 - \frac{w}{\sqrt{3}}\right)^2}.$$

Consequently, $f_1(w) = -(3\sqrt{3}/2)T(w)T'(w)$ and $F_1(w) = -(3\sqrt{3}/4)[T(w)^2 - (1/3)]$. Define $G_1 = F_1 \circ S$. Then $\|F_1\| = \|G_1\|$, $G_1(z) = -(3\sqrt{3}/4)[z^2 - (1/3)]$ and

$$(1 - |z|^2)|G_1'(z)| = \frac{3\sqrt{3}}{2}(1 - |z|^2)|z|.$$

The function $h(t) = (1 - t^2)t$ vanishes at $t = 0, 1$ and attains its maximum value of $2/3\sqrt{3}$ on the interval $[0, 1]$ uniquely at the point $t = 1/\sqrt{3}$. Thus, $\|G_1\| = 1$ and $(1 - |z|^2)|G_1'(z)| = 1$ if and only if $|z| = 1/\sqrt{3}$. This implies that $\|F_1\| = 1$ and $(1 - |w|^2)|F_1'(w)| = 1$ if and only if w lies on the circle through 0 and $\sqrt{3}/2$ that is symmetric about the real axis.

Next, we determine $r(0, F_1)$ and $s(0, F_1)$ geometrically. We observe that $r(0, F_1) = r(-1/\sqrt{3}, G_1)$ and $s(0, F_1) = s(-1/\sqrt{3}, G_1)$. Clearly, the function G_1 is a two-sheeted branched covering of $\mathbb{D}$ onto the disk $D(\sqrt{3}/4, 3\sqrt{3}/4)$ with $G_1(0) = \sqrt{3}/4$ and $G_1'(0) = 0$. Thus, the branch point closest to $G_1(-1/\sqrt{3}) = 0$ is $\sqrt{3}/4$, so that $r(-1/\sqrt{3}, G_1) = \sqrt{3}/4$. Also, G_1 is univalent in the half-plane $\{z : \operatorname{Re} z < 0\}$ and $G_1'(0) = 0$, so $s(-1/\sqrt{3}, G_1) = 2 \operatorname{artanh} 1/\sqrt{3}$, the hyperbolic distance between $-1/\sqrt{3}$ and the origin.

We also give analytic proofs of the facts $r(0, F_1) = \sqrt{3}/4$ and $s(0, F_1) = 2 \operatorname{artanh} 1\sqrt{3}$. Since

$$\begin{aligned} \operatorname{Re} F_1'(w) = \operatorname{Re} f_1(w) &= 1 - \sum_{n=2}^{\infty} \frac{(n^2 - 1)\operatorname{Re}(w^n)}{(\sqrt{3})^n} \\ &\geq 1 - \sum_{n=2}^{\infty} \frac{(n^2 - 1)|w^n|}{(\sqrt{3})^n} \\ &= f_1(|w|) = F_1'(|w|) > 0 \quad \text{if } |w| < 1/\sqrt{3} \end{aligned}$$

and $F_1'(1/\sqrt{3}) = f_1(1/\sqrt{3}) = 0$, the Wolff-Warschawski-Noshiro Theorem shows that F_1 is univalent in $D(1/\sqrt{3})$ but in no larger disk. Thus, $s(0, F_1) = 1/\sqrt{3}$. Next, we show that F_1 maps $D(1/\sqrt{3})$ conformally onto a region that contains the disk $D(\sqrt{3}/4)$. For $w = e^{i\theta}/\sqrt{3}$ we have

$$F_1(w) = \int_0^w F_1'(\omega)d\omega = \int_0^{1/\sqrt{3}} F_1'(re^{i\theta})ie^{i\theta}dr,$$

so that

$$|F_1(w)| \geq \operatorname{Re}\frac{F_1(w)}{ie^{i\theta}} = \int_0^{1/\sqrt{3}} \operatorname{Re}F_1'(re^{i\theta})dr \geq \int_0^{1/\sqrt{3}} F_1'(r)dr = F_1(1/\sqrt{3}) = \sqrt{3}/4.$$

Thus, $F_1(D(1/\sqrt{3})) \supset D(\sqrt{3}/4)$. Since $F_1(1/\sqrt{3}) = \sqrt{3}/4$ and $F_1'(1/\sqrt{3}) = 0$, it follows that $r(0, F_1) = \sqrt{3}/4$.

It is not difficult to show that $r(F) = 3\sqrt{3}/8$ and $s(F) = +\infty$.

Theorem 1. *Suppose that $s > 0$, $|a| = s$ and g is holomorphic and nonconstant on $D(s) \cup \{a\}$. If $|g(z)| \le |g(a)|$ for $z \in D(s)$, then $c = ag'(a)/g(a)$ is positive and*

$$Re\left\{\frac{g(ta)}{g(a)}\right\} \ge \frac{(c+1)t-(c-1)}{(c+1)-(c-1)t}, \quad t \in (-1,1),$$

with equality for some $t \in (-1,1)$ if and only if

$$g(z) = g(a)\frac{(c+1)z-(c-1)a}{(c+1)a-(c-1)z}.$$

Proof. We start by reducing the general case to the special case in which $a = 1$ and $g(a) = 1$. Set $h(z) = g(az)/g(a)$. Then h is holomorphic in $\mathbb{D} \cup \{1\}$, $h(\mathbb{D}) \subset \mathbb{D}, h(1) = 1$ and $c = h'(1)$. Since $h(\mathbb{D}) \subset \mathbb{D}$ and $h(1) = 1$, elementary geometric considerations show that $c > 0$. Set

$$U(z) = \frac{z - \dfrac{c-1}{c+1}}{1 - \dfrac{c-1}{c+1}z} = \frac{(c+1)z-(c-1)}{(c+1)-(c-1)z}.$$

Note that U is a conformal automorphism of $\mathbb{D}$, $U(1) = 1$ and $U'(1) = c$. We need to prove that $\mathrm{Re}h(t) \ge U(t)$, $t \in (-1,1)$, with equality if and only if $h = U$.

Set $k = U^{-1} \circ h$. Then k is holomorphic in $\mathbb{D} \cup \{1\}$, $k(\mathbb{D}) \subset \mathbb{D}$, $k(1) = 1$ and $1 = k'(1)$. Recall that for $r \in (0,\infty)$

$$\Delta(1,r) = \{z \in \mathbb{D} : \frac{|1-z|^2}{1-|z|^2} < r\} = D(\frac{1}{1+r}, \frac{r}{1+r})$$

is a horodisk in $\mathbb{D}$ based at 1; that is, an open disk in $\mathbb{D}$ that is internally tangent to the unit circle at 1. Observe that $U(\Delta(1,r)) = \Delta(1,cr)$. Julia's Lemma ([A$_2$, pp. 7-9], [C, pp. 23-28]) asserts that for $z \in \mathbb{D}$

$$\frac{|1-k(z)|^2}{1-|k(z)|^2} \le |k'(1)|\frac{|1-z|^2}{1-|z|^2} = \frac{|1-z|^2}{1-|z|^2},$$

with equality for some $z \in \mathbb{D}$ if and only if $k = R_b$ for some $b \in \mathbb{R}$, where

$$R_b(z) = \frac{(1+z)-(1+ib)(1-z)}{(1+z)+(1-ib)(1-z)}.$$

Note that R_b is a conformal automorphism of $\mathbb{D}$ that fixes 1 and maps each horodisk $\Delta(1,r)$ onto itself. Julia's Lemma has an elegant geometric interpretation: either $k = R_b$ for some $b \in \mathbb{R}$ or else $k(\overline{\Delta(1,r)}\backslash\{1\}) \subset \Delta(1,r)$ for all $r \in (0,\infty)$. Thus, either $h = U \circ R_b$ for some $b \in \mathbb{R}$ or else $h(\overline{\Delta(1,r)}\backslash\{1\}) \subset U(\Delta(1,r))$ for all $r \in (0,\infty)$.

Given $t \in (-1,1)$, select $r = (1-t)/(1+t) \in (0,\infty)$. Then $\Delta(1,r)$ is the horodisk in $\mathbb{D}$ whose boundary meets the real axis in the points t and 1. From the inclusion $h(\Delta(1,r)) \subset U(\Delta(1,r))$ we obtain $\mathrm{Re}h(t) \ge U(t)$ with equality for some t if and only if

$\mathrm{Im}h(t) = 0$ and $h(t) = U(t)$, that is, $k(t) = t$. But then equality holds in Julia's Lemma, so $k = R_b$ for some $b \in \mathbf{R}$. But if $t = R_b(t)$ for some $t \in (1-, 1)$, then it follows that $b = 0$. Since R_0 is the identity function, we deduce $h = U$.

Corollary. *Suppose $s > 0$, $|a| = s$ and g is holomorphic and nonconstant on $D(s) \cup \{a\}$. If $|g(z)| \le |g(a)|$ for $z \in D(s)$ and $ag'(a) = g(a)$, then*

$$\mathrm{Re}\left\{\frac{g(ta)}{g(a)}\right\} \ge t, \quad t \in (-1,1),$$

with equality for some $t \in (-1,1)$ if and only if $g(z) = g(a)z/a$.

Remark. For the proof of Theorem 1 we only require the weak form of Julia's Lemma in which g is assumed to be analytic at $z = 1$, not the general result dealing with the angular derivative at $z = 1$. There is a simple proof of Julia's Lemma in this special case in [PS, prob. 292, p. 141].

Now, we give a geometric proof of Bonk's distortion theorem with a careful analysis of the sharpness.

Theorem 2. *Suppose $F \in \mathcal{B}, \|F\| = 1$ and $F'(0) = 1$. Then $\mathrm{Re}F'(w) \ge F_1'(|w|)$ for $|w| \le \sqrt{3}/2$ with equality at $re^{i\theta} \ne 0$ if and only if $F(w) = e^{i\theta}F_1(e^{-i\theta}w)$.*

Proof. There is no harm in assuming that $w = u \in (0,1)$; if not, then consider $e^{-i\theta}F(e^{i\theta}w)$ to treat the case in which the point is $re^{i\theta}$. Note that

$$|1 + F''(0)w + \cdots| = |F'(w)| \le \frac{1}{1-|w|^2} = 1 + |w|^2 + \cdots$$

implies $F''(0) = 0$.

Consider any $a, 0 < a < 1$. Set $w = S_a(z) = (z+a)/(1+az)$ and $G_a = F \circ S_a$. Then $\|F\| = \|G_a\|$ and $(1-|z|^2)|G_a'(z)| = 1$ for $z = -a$ since $S_a(-a) = 0$ and $F'(0) = 1$. Thus, for $|z| \le a$ we have

$$|G_a'(z)| \le \frac{1}{1-|z|^2} \le \frac{1}{1-a^2} = |G_a'(-a)|.$$

In fact, direct calculation reveals $G_a'(-a) = 1/(1-a^2)$ and $G_a''(-a) = -2a/(1-a^2)^2$, so that

$$c = -\frac{aG_a''(-a)}{G_a'(-a)} = \frac{2a^2}{1-a^2}.$$

The preceding theorem applied to $g = G_a'$ yields

$$\mathrm{Re}\frac{G_a'(-at)}{G_a'(-a)} \ge \frac{(c+1)t-(c-1)}{(c+1)-(c-1)t} = \frac{(1-3a^2)+(1+a^2)t}{(1+a^2)+(1-3a^2)t}, \quad t \in (-1,1),$$

or

$$\mathrm{Re}\{F'(S_a(-at))S_a'(-at)\} \ge \frac{(1-3a^2)+(1+a^2)t}{(1-a^2)[(1+a^2)+(1-3a^2)t]}, \quad t \in (-1,1),$$

with equality for some $t \in (-1,1)$ if and only if

$$G_a'(z) = \frac{1}{1-a^2}\frac{(c+1)z-(c-1)(-a)}{(c+1)(-a)-(c-1)z}.$$

Set $u = S_a(-at)$, or $t = (a-u)/a(1-au)$. Observe that S_a maps the interval $(-a, a)$ onto the interval $(0, 2a/(1+a^2))$ and that

$$S_a'(-at) = \frac{1}{(S_a^{-1})'(u)} = \frac{(1-au)^2}{1-a^2}.$$

Therefore,

$$(*) \qquad \mathrm{Re}F'(u) \geq \frac{2a-(1+3a^2)u}{(1-au)^2[2a-(1-a^2)u]}, \quad u \in (0, 2a/(1+a^2)).$$

This implies that

$$\mathrm{Re}F'(w) \geq \frac{2a-(1+3a^2)|w|}{(1-a|w|)^2[2a-(1-a^2)|w|]}, \quad |w| < 2a/(1+a^2).$$

Thus, $\mathrm{Re}F'(w) > 0$ for $|w| < 2a/(1+3a^2)$. The choice $a = 1/\sqrt{3}$ produces the largest disk on which $\mathrm{Re}F'(w) > 0$ and also yields $\mathrm{Re}F'(w) \geq F_1'(|w|)$ for $|w| \leq \sqrt{3}/2$. Note that for $a = 1/\sqrt{3}$ we have $c = 1$ and so strict inequality holds in (*) for $u \in (0, \sqrt{3}/2)$ unless $G_a'(z) = -3\sqrt{3}z/2$, or equivalently, $G_a(z) = G_1(z)$, and so $F(w) = F_1(w)$. Here G_1 refers to the function in Example 1. Conversely, if $F = F_1$, then for $a = 1/\sqrt{3}$ equality holds in (*) at each point of the interval $(0, \sqrt{3}/2)$.

Corollary 1. *Suppose $F \in \mathcal{B}$, $\|F\| = 1$ and $F'(0) = 1$. Then $r(0, F) \geq \sqrt{3}/4$ and $s(0, F) \geq 2\,\mathrm{artanh}\,1/\sqrt{3}$ with strict inequality unless $F(w) = e^{i\theta}F_1(e^{-i\theta}w)$ for some $\theta \in \mathbf{R}$.*

Proof. Suppose $F(w) \neq e^{i\theta}F_1(e^{-i\theta}w)$ for all $\theta \in \mathbf{R}$. Then $\mathrm{Re}F'(w) > F_1'(|w|) > 0$ when $|w| < \sqrt{3}/2$. In particular, $\min\{\mathrm{Re}F'(w) : |w| = 1/\sqrt{3}\} > 0$. Thus, there exists $s > 1/\sqrt{3}$ such that $\mathrm{Re}F'(w) > 0$ on $D(s)$. The Wolff-Warschawski-Noshiro Theorem implies that F is univalent on $D(s)$, so $s(0, F) \geq 2\,\mathrm{artanh}\, s > 2\,\mathrm{artanh}\,1/\sqrt{3}$. Also, for $w = e^{i\theta}/\sqrt{3}$ we have

$$|F(w)| \geq \mathrm{Re}\left\{\frac{F(w)}{ie^{i\theta}}\right\} = \int_0^{1/\sqrt{3}} \mathrm{Re}F'(re^{i\theta})dr > \int_0^{1\sqrt{3}} F_1'(r)dr = \frac{\sqrt{3}}{4}.$$

Hence, $\min\{|F(w)| : |w| = 1/\sqrt{3}\} > \sqrt{3}/4$, so there exists $r > \sqrt{3}/4$ such that $F(D(1/\sqrt{3}))$ contains $D(r)$. Because F is univalent in $D(1/\sqrt{3})$, this implies that $r(0, F) > \sqrt{3}/4$. ■

Remark. For bounded analytic functions there is a sharp analogy of Corollary 1 due to Landau, see ([H_2, pp, 36-39]). The extremal functions for Landau's result are two-sheeted branched coverings of $\mathbf{D}$ onto itself. This result of Landau is the basis for one elementary proof of a lower bound for the Bloch constant, see ([H_2, pp. 46]).

Corollary 2. $B > \sqrt{3}/4$ *and* $M > 2\,\mathrm{artanh}\,1/\sqrt{3}$.

Proof. Let $\mathcal{B}_1 = \{F : F \in \mathcal{B}, \|F\| = 1 \text{ and } F'(0) = 1\}$. The family $\mathcal{B}_1$ is a compact normal family. Robinson [R] showed that there exist extremal functions for the Bloch constant, that is, functions $F \in \mathcal{B}_1$ with $r(F) = B$. Consider any such extremal function F. If $F(w) \neq e^{i\theta}F_1(e^{-i\theta}w)$ for all $\theta \in \mathbf{R}$, then $r(F) > r(0, F) > \sqrt{3}/4$ by the preceding corollary. On the other hand, if $F(w) = e^{i\theta}F_1(e^{-i\theta}w)$ for some $\theta \in \mathbf{R}$, then $r(F) = 3\sqrt{3}/8$ by Example 1. Thus, in either case, $B > \sqrt{3}/4$.

Next, we show $M > 2\,\text{artanh}\,1/\sqrt{3}$ by a similar argument. Consider any sequence $\{F_n\}$ in $\mathcal{B}_1$ with $s(F_n) \to M$. Because $\mathcal{B}_1$ is a compact normal family, we may assume that $F_n \to F \in \mathcal{B}_1$, where the convergence is uniform on compact subsets of $\mathbf{D}$. If F is univalent on $D_h(a, s)$, then for each $\varepsilon > 0$ there exists N such that for all $n \geq N$, F_n is univalent on $D_h(a, s-\varepsilon)$. Hence, for $n \geq N$, $s(F_n) \geq s(a, F_n) \geq s(a, F) - \varepsilon$. By letting n tend to infinity we obtain $M \geq s(a, F) - \varepsilon$. But $\varepsilon > 0$ is arbitrary, so $M \geq s(a, F)$. This yields $M \geq s(F)$, so $M = s(F)$. This demostrates that there are extremal functions for the Marden constant. The remainder of the proof that $M > 2\,\text{artanh}\,1/\sqrt{3}$ is now analogous to the proof in the preceding paragraph and is omitted. ■

3. Geometric interpretation

Suppose F is holomorphic and nonconstant on $\mathbf{D}$. Let $X = X(F)$ be the Riemann image surface of F viewed as spread over the complex plane $\mathbf{C}$. In connection with the Bloch constant problem, it is natural to inquire about the possible location of the largest schlicht disk on X. One plausible guess is that the largest schlicht disk would be centered at a point where the hyperbolic metric attains its minimum value. For a plane region it makes sense to speak of the minimum of the density of the hyperbolic metric, but for a Riemann surface it makes no sense to speak of the value of a metric at a point of the surface. However, for a Riemann surface spread over $\mathbf{C}$ there is a natural way to define a density for the hyperbolic metric. This density will be infinite at all branch points and at all finite boundary points. We will show that a minimum point for this density is always the center of a relatively large schlicht disk, but generally not the largest such disk on the surface. In fact, the work of Bonk yields a sharp lower bound for the radius of a schlicht disk centered at such a minimum point; this is the geometric significance of $\sqrt{3}/4$. The result of Bonk also gives the best possible lower bound on the hyperbolic radius of a hyperbolic disk centered at a minimum point which is one-sheeted when viewed as spread over $\mathbf{C}$. This is a geometric interpretation of $2\,\text{artanh}\,1/\sqrt{3}$. This is the same as the lower bound on the hyperbolic distance from a minimum point to the nearest branch point.

We begin by making precise the notion of the Riemann image surface and establish some notation. Precisely, $X = \{(w, F(w)) : w \in \mathbf{D}\}$ is the graph of F in $\mathbf{C} \times \mathbf{C}$. Define the two natural projections of X onto the coordinates planes:

$$\begin{aligned} &\pi_1 : X \to \mathbf{D} && \pi_2 : X \to \mathbf{C} \\ &\pi_1(w, F(w)) = w, && \pi_2(w, F(w)) = F(w). \end{aligned}$$

The surface X is endowed with the unique conformal structure that makes both π_1 and π_2 analytic functions. Note that π_1 is a conformal mapping of the simply connected

Riemann surface X onto $\mathbf{D}$. The set of branch points of X is $b(X) = \{(w, F(w)) : F'(w) = 0\}$. The function π_2 can be used as a local coordinate at each point of $X \backslash b(X)$. For convenience, let

$$\pi_1^{-1} = \tilde{F} : \mathbf{D} \to X,$$
$$\tilde{F}(w) = (w, F(w)),$$

and observe that $\pi_2 \circ \tilde{F} = F$.

There are two natural metrics and associated distance functions on X. First, there is the hyperbolic metric $\lambda_X(\omega)|d\omega|$ and the associated distance function d_X. Recall that the hyperbolic metric on X is the unique conformal metric on X whose pull-back under any conformal mapping of $\mathbf{D}$ onto X is the hyperbolic metric on $\mathbf{D}$; in symbols,

$$\tilde{F}^*(\lambda_X(\omega)|d\omega|) = \lambda_{\mathbf{D}}(w)|dw| = \frac{2|dw|}{1 - |w|^2};$$

where $\lambda_{\mathbf{D}}(w)|dw|$ is the ordinary hyperbolic metric on $\mathbf{D}$ with curvature -1. Second, there is the pull-back of the euclidean metric via π_2. Explicitely, $\lambda_{\mathbf{C}}(\zeta)|d\zeta| = 1|d\zeta|$ is the euclidean metric on $\mathbf{C}$ and $\pi_2^*(\lambda_{\mathbf{C}}(\zeta)|d\zeta|)$ is a metric on $X \backslash b(X)$, but just a semi-metric on X itself. In other words, $\pi_2^*(\lambda_{\mathbf{C}}(\zeta)|d\zeta|)$ is a continuous, nonnegative linear density on X which vanishes precisely at the branch points. Because the zeros of this semi-metric are isolated points on X, it induces a distance function d on X that is compatible with the topology of X.

Note that $r(a, F)$ is the minimum of the distance from $(a, F(a))$ to $b(X)$ and the distance from $(a, F(a))$ to the ideal boundary of X relative to the distance function d. Crudely speaking, it is the radius of the largest disk on X, relative to the distance function d, centered at $(a, F(a))$ which does not contain a branch point or meet the ideal boundary of X. Similarly, let $t(a, F)$ be the minimum of the distance from $(a, F(a))$ to $b(X)$ and the distance from $(a, F(a))$ to the ideal boundary of X relative to the distance function d_X. In fact, the latter distance is always infinite, so $t(a, F)$ is just the d_X-distance from $(a, F(a))$ to the set of branch points. The geometric quantity $t(a, F)$ was introduced and studied in [M$_2$]. Also, observe that $s(a, F)$ is the radius (relative to d_X) of the largest disk centered at $(a, F(a))$ in which π_2 is univalent. In other words, it is the radius of the largest hyperbolic disk centered at $(a, F(a))$ which is one-sheeted when viewed as spread over the complex plane. Clearly, $s(a, F) \leq t(a, F)$.

The quotient

$$\mu = \mu_X = \frac{\lambda_X(\omega)|d\omega|}{\pi_2^*(\lambda_{\mathbf{C}}(\zeta)|d\zeta|)}$$

of the hyperbolic metric by the pull-back of the euclidean metric defines a positive, continuous function on X which is infinite at each branch point. The function μ is the density of the hyperbolic metric when the hyperbolic metric is expressed in terms of the local parameter π_2. Observe that

$$\mu(w, F(w)) = \mu(\tilde{F}(w)) = \frac{\tilde{F}^*(\lambda_X(\omega)|d\omega|)}{\tilde{F}^*(\pi_2^*(\lambda_{\mathbf{C}}(\zeta)|d\zeta|))} =$$

$$= \frac{\tilde{F}^*(\lambda_X(\omega)|d\omega|)}{F^*(\lambda_{\mathbf{C}}(\zeta)|d\zeta|)} = \frac{\lambda_{\mathbf{D}}(w)|dw|}{|F'(w)||dw|} = \frac{2}{(1-|w|^2)|F'(w)|}.$$

Here we have used $F^* = (\pi_2 \circ \tilde{F})^* = \tilde{F}^* \circ \pi_2^*$ and $\tilde{F}^*(\lambda_X(\omega)|d\omega|) = \lambda_{\mathbf{D}}(w)|dw|$. Set

$$m = m(X) = \inf\{\mu(w, F(w)) : (w, F(w)) \in X\}.$$

From the preceding work it is clear that $m(X) = 2/\|F\|$. Also, μ attains its minimum value m at the point $(a, F(a))$ if and only if $(1-|a|^2)|F'(a)| = \|F\|$. We can now give a geometric version of Corollary 1 of Theorem 2.

Theorem 3. *Suppose F is holomorphic in $\mathbf{D}$ and $(a, F(a))$ is a minimum point for μ_X. Then $r(a,F)m(X) \geq \sqrt{3}/4$ and $t(a,F) \geq s(a,F) \geq 2\,\mathrm{artanh}\, 1/\sqrt{3}$. Both of these inequalities are sharp.*

Proof. We may assume that $a = 0$ and $F'(0) > 0$. If not, then replace F by $\lambda F \circ S$, where λ is a unimodular constant and $S(z) = (z+a)/(1+\bar{a}z)$. Notice that we have $m(X(F)) = m(X(\lambda F \circ S))$. Thus, we are assuming $F'(0) = \|F\|$. Then the function $F/\|F\|$ satisfies the hypotheses of Corollary 1 to Theorem 2, and so the conclusions of Theorem 3 follow immediately. Note that equality holds in all the inequalities for the function F_1. ∎

Example 2. It is interesting to look at this geometric interpretation in the special case of the function F_1. The extremal function F_1 has an intimate connection with the ultrahyperbolic metric used in the proof that $B > \sqrt{3}/4$ via Ahlfors' method. This sheds light upon why Ahlfors' method and Bonk's distortion theorem yield the same lower bound for the Bloch constant.

Actually, it is simpler to consider the function $F_0 = F_1 - \sqrt{3}/4$. Then F_0 is a two-sheeted branched covering of $\mathbf{D}$ onto $D(3\sqrt{3}/4)$ and we let X_0 denote the Riemann image surface of F_0. The associated function is $G_0(z) = F_0 \circ S(z) = -(3\sqrt{3}/4)z^2$. The function μ satisfies

$$\mu(w, F_0(w)) = \frac{2}{(1-|w|^2)|F_0'(w)|} = \frac{2}{(1-|z|^2)|G_0'(z)|}.$$

If $F_0(w_1) = F_0(w_2)$, then $G_0(T(w_1)) = G_0(T(w_2))$, so $T(w_1) = -T(w_2)$. This implies that $G_0'(T(w_1)) = -G_0'(T(w_2))$, so $\mu(w_1, F_0(w_1)) = \mu(w_2, F_0(w_2))$. In brief, the function μ is invariant under the sheet-interchange function for X_0, so μ induces a function σ on $D(3\sqrt{3}/4)$. We explicitely determine σ. If $\zeta = G_0(z) = F_0(w)$, then

$$\sigma(G_0(z)) = \sigma(F_0(w)) = \sigma(\pi_2(w, F_0(w))) = \mu(w, F_0(w)) = \frac{2}{(1-|z|^2)|G_0'(z)|}.$$

For $\zeta = G_0(z) = -(3\sqrt{3}/4)z^2$ this gives

$$\sigma(\zeta) = \frac{\sqrt{\dfrac{3\sqrt{3}}{4}}}{|\zeta|^{1/2}\left(\dfrac{3\sqrt{3}}{4} - |\zeta|\right)}.$$

$\sigma(\zeta)|d\zeta|$ is a conformal metric on the punctured disk $D(3\sqrt{3}/4)\backslash\{0\}$ with constant Gaussian curvature -1.

Now, the connection with Ahlfors' method becomes apparent. We employ the notation of [M_1, §5]. In Ahlfors' method the appropiate ultrahyperbolic metric is constructed from the conformal metric

$$\sigma_{1,\tau}(\zeta)|d\zeta| = \frac{\sqrt{3\tau}|d\zeta|}{|\zeta|^{1/2}(3\tau - |\zeta|)}.$$

(Actually, this metric is twice the metric that appears in [M_1, §5]; the reason for this factor of two is that $\lambda_{\mathbf{C}}(\zeta)|d\zeta|$ was taken to be twice the euclidean metric in [5].) This is a conformal metric on $D(3\tau)\backslash\{0\}$ which has constant Gaussian curvature -1. The proof that $B > \sqrt{3}/4$ via Ahlfors' method uses the value $\tau = \sqrt{3}/4$. But for this choice of τ, $\sigma_{1,\tau} = \sigma$.

References

[A_1] L.V. Ahlfors, *An extension of Schwarz's lemma*, Trans. Amer. Math. Soc. **43** (1938), 359-364.

[A_2] L.V. Ahlfors, *Conformal invariants: Topics in geometric function theory*, McGraw Hill, New York, 1973.

[AG] L.V. Ahlfors, H. Grunsky, *Über die Blochsche Konstante*, Math. Z. **42** (1937), 671-673.

[B] M. Bonk, *On Bloch's constant*, Proc. Amer. Math. Soc. (to appear).

[C] C. Carathéodory, *Theory of functions of a complex variable*, vol. II, 2nd English ed., Chelsea Publishing Co., New York, 1960.

[H_1] M. Heins, *On a class of conformal metrics*, Nagoya Math. J. **21** (1962), 1-60.

[H_2] M. Heins, *Selected topics in the classical theory of functions of a complex variable*, Holt, Rinehart and Winston, New York, 1962.

[L_1] E. Landau, *Der Picard-Schottkysche Satz und die Blochsche Konstante*, Sitzungsber. Preuss. Akad. Wiss. Berlin, Phys.-Math. Kl. **32** (1926), 467-474.

[L_2] E. Landau, *Über die Blochsche Konstante und zwei verwandte Weltkonstanten*, Math. Z. **30** (1929), 608-634.

[M_1] D. Minda, *Bloch constants*, J. Analyse Math. **41** (1982), 54-84.

[M_2] D. Minda, *Domain Bloch constants*, Trans. Amer. Math. Soc. **276** (1983), 645-655.

[M_3] D. Minda, *Marden constants for Bloch and normal functions*. J. Analyse Math. **42** (1982/83), 117-127.

[Pe_1] E. Peschl, *Über die Verwendung von Differentialinvarianten bei gewissen Funktionenfamilien und die Übertragung einer darauf gegründeten Methode auf partielle Differentialgleichungen vom elliptischen Typus*, Ann. Acad. Sci. Fenn. Ser. AI **336/6** (1963), 23 pp.

[Pe_2] E. Peschl, *Über unverzweigte konforme Abbildungen*, Österreich. Akad. Wiss. Math.-Naturwiss. Kl. S.-B. II **185** (1976), 55-78.

[PS] G. Pólya, G. Szegö, *Aufgaben und Lehrsätze aus der Analysis*, Bd. 1, Die Grundlehren der math. Wissenschaften in Einzeldarstellungen, Bd. 19, Springer-Verlag, New York, 1964.

[Po] Ch. Pommerenke, *On Bloch functions*, J. London Math. Soc. (2) **2** (1970), 689-695.

[R] R.M. Robinson, *Bloch functions*, Duke Math. J. **2** (1936), 453-459.

Received: March 18, 1989

Computational Methods and Function Theory
Proceedings, Valparaíso 1989
St. Ruscheweyh, E.B. Saff, L. C. Salinas, R.S. Varga (*eds.*)
Lecture Notes in Mathematics **1435**, pp. 143–154

On some analytic and computational aspects of two dimensional vortex sheet evolution

O. F. Orellana[1]

Departamento de Matemáticas
Universidad Técnica Federico Santa María
Valparaíso, Chile

Abstract. This survey paper gives an account of recent analytic and numerical results of the initial value problem:

$$\frac{\partial \overline{Z}}{\partial t}(\gamma,t) = \frac{1}{2\pi i}\int_{-\infty}^{\infty} \frac{d\gamma'}{Z(\gamma,t)-Z(\gamma',t)},$$

$$Z(\gamma,0) = \gamma + S(\gamma),$$

which is the Birkhoff-Rott equation for the evolution of a sligthly perturbed flat vortex sheet. We will indicate some open problems of current research and propose a new physically desingularized Vortex sheet equation, which agrees with the finite thickness vortex layer equations in the localized approximation.

Introduction

There exist two main motivations to study the problem we are concerned about in these notes: in three dimensions, it is an important unsolved problem of mathematical fluid dynamics to determine whether solutions of the Euler equations develop singularities in finite time, with smooth initial data. This problem is important for two reasons: from the physical point of view because the existence of such singularities is connected with the onset of turbulence in a high Reynolds number regime, and, from the numerical point of view because the existence of such singularities will generate instabilities in the calculations. Since, this seems to be a hard problem that has been studied through different numerical methods by several authors it is not a bad idea to try a simpler problem namely: show that initially singular solutions can become more singular after a finite

[1]The author was economically supported by FONDECYT (grant number 235, 1987–88) and Universidad Técnica Federico Santa María (grants 88.12.08, 1988 and 89.12.08) 1989).

time. Specifically, suppose the velocity field has a symmetry with respect to one of the axes so that the velocity field is two dimensional and has a discontinuity along a smooth curve (for instance a vortex sheet). Does the initially analytic curve stay analytic for all time or develop a singularity in finite time? A number of authors have shown through analytic and/or numerical techniques the appearance of a singularity in finite time (see [1], [2], [3], [4], [5], [7]).

The second motivation to study the Birkhoff-Rott equation as a mathematical model for the evolution of a vortex sheet is because of its applications to aerodynamics and consequently to the design of more secure and economic airplanes through the calculation of lift of airfoils, the design of airfoils and knowledge and control of the region of turbulence (see [8]).

A mathematically rigorous deduction of the Birkhoff-Rott equation can be found in [9] and [10].

This review article describes mathematical analysis results about the evolution of a slightly perturbed flat vortex sheet, because of its importance to the numerics and computation of vortex sheet evolution. Hence we will also comment on the numerical methods to solve Birkhoff-Rott equation proposed by R. Krasny (see [5] and [6]) and the proof of convergence of such methods given by R. Caflisch and J. Lowengrub in [17]. We also mention open questions of current research in the appropiate place.

In the last section of this paper we comment on a new physically desingularized vortex sheet equation proposed by Caflisch, Orellana and Siegel [14] as an alternative to the desingularized equation proposed by R. Krasny [6].

1. Instability analysis and singularity formation for vortex sheets.

We will call a 2–dimensional vortex sheet, a discontinuity curve in a fluid domain, which moves with velocity equal to the average of the velocities on its two sides, and across which the tangential velocity, but not the normal velocity, is discontinuous. The jump in tangential velocity across the sheet in a given point is called vortex sheet strength. A good example is provided by the flow field with velocity components:

$$(u,v)(x,y) := \begin{cases} (-\frac{1}{2},0) & \text{if } y > 0 \\ (+\frac{1}{2},0) & \text{if } y < 0 \end{cases} \tag{1}$$

which is a steady weak solution of Euler equations. The flat interface $F := \{(x,y) \in \mathbf{R}^2 : y = 0\}$ defines a vortex sheet of uniform strength equal to one, which moves with velocity equal to zero.

If we represent the vortex sheet and its corresponding curve in the complex plane by

$$Z(\gamma,t) = X(\gamma,t) + iY(\gamma,t),$$

where t is the time variable and γ is the Lagrangian parameter which measure the circulation between a reference point ($\gamma = 0$) and an arbitrary point on the sheet, then the evolution of the vortex sheet is governed by the singular-integrodifferential equation:

$$\frac{\partial \overline{Z}}{\partial t}(\gamma, t) = \frac{1}{2\pi i} \fint_{-\infty}^{\infty} \frac{d\gamma'}{Z(\gamma, t) - Z(\gamma', t)} \tag{2}$$

where the bar on the left side denotes complex conjugate and the slash on the integral sign denotes Cauchy principal value due to the singularity of the integrand at $\gamma = \gamma'$. Moreover the vortex sheet strength $\sigma(\gamma, t)$ is the jump in tangential velocity across the sheet at $Z(\gamma, t)$ and it is determined up to sign by:

$$\sigma(\gamma, t) = \left| \frac{\partial Z}{\partial \gamma} \right|^{-1} \tag{3}$$

Moreover $Z = \gamma$ defines a flat vortex sheet of uniform strength equal to one which is the steady solution of (2) corresponding to the weak steady solution (1) of Euler equations.

Equation (2) was derived using the Biot-Savart law by Birkhoff [9] and in a more mathematically rigorous way by Sulem, Sulem, Bardos and Frisch [10]. Therefore the nonlinear evolution of a vortex sheet with a small initial perturbation is given by the initial value problem:

$$\begin{aligned} \frac{\partial \overline{Z}}{\partial t}(\gamma, t) &= \frac{1}{2\pi i} \int_{-\infty}^{\infty} \frac{d\gamma'}{Z(\gamma, t) - Z(\gamma', t)} \\ Z(\gamma, 0) &= \gamma + S(\gamma, 0) \end{aligned} \tag{4}$$

It is well known that the flat 2-d vortex sheet $Z(\gamma, t) = \gamma$ is linearly unstable to small, analytic disturbances.

When a slight disturbance preserving the irrotationality of the flow outside the interfase is considered, a linear analysis gives as general modes

$$\begin{aligned} Z(\gamma, t) &= \gamma + (1 - i)\varepsilon e^{\frac{n}{2}t}\{\alpha_1 \cos n\gamma - \beta_1 \sin n\gamma\} \\ Z(\gamma, t) &= \gamma + (1 + i)\varepsilon e^{-\frac{n}{2}t}\{\beta_2 \cos n\gamma - \alpha_2 \sin n\gamma\} \end{aligned} \tag{5}$$

This shows that the amplitude of disturbances of wavenumber n may grow exponentially in time at the rate $|\frac{n}{2}|$; this phenomenon is known by the name of Kelvin-Helmholtz instability (see Batchelor [11]) and since the growth rate is unbounded the linear problem is ill-posed in the sense of Hadamard (see Garabedian [12]).

From the physical point of view the flow is made well posed by the inclusion of viscosity or non-zero thickness for the Vortex sheet. Mathematically the problem may be made well-posed by considering the solution in the class of analytic function of γ.

The initial value problem (4) and (5) has been analyzed by Moore [1] and [2]. He uses a novel formal perturbation analysis to obtain a simpler equation which approximates the singular integral equation (4). Then in [2] performing changes of variables and transformation, Moore obtains a 2×2 nonlinear hyperbolic system in the independent variable t and y, where y is the real part of $i\gamma$ after the complexification of γ (i.e. Moore's nonlinear hyperbolic system lives on a plane perpendicular to the physical plane). Finally, Moore performs an asymptotic analysis on his nonlinear hyperbolic system to describe the formation of a singularity at a critical time t_c. For an initial perturbation $S(\gamma, 0) = i\varepsilon \sin\gamma$, he showed formally that one of the family of characteristics of his conservation laws has an envelope in the domain $D = \{(y,t) : t > 0\}$ on which the solution has a singularity. This envelope has the equation

$$y = \ln\left\{\frac{4}{\varepsilon(t + e^{-t} - 1)}\right\} - \frac{t}{2} - 1.$$

Therefore for small t and $y >> 1$, the singularity is far from the physical plane. As t increases it moves towards the physical plane and reaches it at the critical time

$$t_c = 2|\log\varepsilon| + 0(\log|\log\varepsilon|)$$

At the singularity the curvature of the vortex sheet is infinite, although the slope of the sheet remains finite and continuous. Moore also predicted that the singularity would be of type $S = \gamma^{3/2}$. For t near t_c it is not clear whether Moore's approximate solution is asymptotically correct. But his results have been partially confirmed numerically by Krasny [5] and Meiron et al [3].

Moreover, numerical computations by Krasny [6] using the vortex blob method show that roll up occurs immediately (or at least very soon) after the first singularity time and experiments indicate that the vortex sheet will roll-up into a tightly wound spiral with two branches.

Under the assumption of analyticity, Caflisch and Orellana [4] using a localized approximation method (see Caflisch, Orellana and Siegel [14]) prove existence almost up to the time of expected singularity formation of the solution of (4)-(5) for a small amplitude, odd, periodic perturbation of the flat vortex sheet. Considering $Z(\gamma,t) = \gamma + S(\gamma,t)$, extending $S(\gamma,t)$ analytically to the complex γ-plane, define $S^*(\gamma) = \overline{S(\overline{\gamma})}$)and write

$$S(\gamma,t) = S_-(\gamma,t) + S_+(\gamma,t)$$

where

$$S_-(\gamma,t) = -\sum_{n=0}^{\infty} A_n(t)e^{-in\gamma}$$

$$S_+(\gamma,t) = \sum_{n=1}^{\infty} A_n(t)e^{in\gamma}$$

because of the oddness of $S(\gamma)$, and under suitable assumptions on $S(\gamma,t)$. We were able to *approximately localize* the equation (4) as follows:

$$\begin{aligned} \frac{\partial S^*}{\partial t}(\gamma,t) &= B[S](\gamma,t) = -\frac{1}{2\pi i}\int_{-\infty}^{\infty} (\xi + s(\gamma+\xi) - s(\gamma))^{-1} d\xi \\ &= B[S_+](\gamma,t) + B[S_-](\gamma,t) + D[S_+, S_-] \end{aligned} \tag{6}$$

This equation is just a definition of D. It is shown in section 4 of [4] that D is small since it depends essentially on product terms S_+S_-.

The first two terms of (6) can be evaluated explicity by contour integration. Since the integral of ξ^{-1} vanishes,

$$B[S_+](\gamma) = \frac{1}{2\pi i}\int_{-\infty}^{\infty} \frac{S_+(\gamma+\xi)-S_+(\gamma)}{\xi + S_+(\gamma+\xi) - S_+(\gamma)}\frac{d\xi}{\xi}$$

$$= \frac{1}{2}\left\{\frac{\partial\gamma S_+}{1+\partial\gamma S_+}\right\}$$

Analogously

$$B[S_-](\gamma) = -\frac{1}{2}\left\{\frac{\partial\gamma S_-}{1+\partial\gamma S_-}\right\},$$

and

$$\frac{\partial}{\partial t}\{S_+^*(\gamma) + S_-^*(\gamma)\} = \frac{1}{2}\left\{\frac{\partial\gamma S_+}{1+\partial\gamma S_+}\right\} - \frac{1}{2}\left\{\frac{\partial\gamma S_-}{1+\partial\gamma S_-}\right\} + D[S_+, S_-],$$

which implies that:

$$\frac{\partial}{\partial t}S_-^*(\gamma) = \frac{1}{2}\left\{\frac{\partial\gamma S_+}{1+\partial\gamma S_+}\right\} + H_+D, \tag{7}$$

$$\frac{\partial}{\partial t}S_+^*(\gamma) = -\frac{1}{2}\left\{\frac{\partial\gamma S_-}{1+\partial\gamma S_-}\right\} + H_-D, \tag{8}$$

where $H_\pm[S] = S_\pm$.

Now consider the change of variables $y = i\gamma$ and the function $f(y) = S_+(\gamma)$. From the oddness of $S(\gamma, t)$ it follows that (7) and (8) are equivalent and $S_+(-\gamma) = -S_-(\gamma)$. Moreover since $\overline{y} = i(-\overline{\gamma})$ it follows that

$$f^*(y) = \overline{f(\overline{y})} = (S_+^*)(-\gamma) = -(S_-^*)(\gamma)$$

Hence neglecting H_+D in (8) we can rewrite it as:

$$\frac{\partial}{\partial t}f^*(y,t) = -\frac{1}{2}\left\{\frac{i\partial yf}{1+i\partial yf}\right\}, \tag{9}$$

which is exactly the equation derived by Moore [1].

Now denoting $\phi = 1 + i\partial yf$ and $\psi = \phi^* = 1 - i\partial yf^*$ and applying derivation and $*$ to (9) we obtain:

$$\begin{aligned} \frac{\partial}{\partial t}\phi &= \frac{i}{2}\frac{\partial}{\partial y}\left(\frac{1}{\psi}\right) \\ \frac{\partial}{\partial t}\psi &= -\frac{i}{2}\frac{\partial}{\partial y}(\frac{1}{\phi}) \end{aligned} \tag{10}$$

which is a system of two conservation laws. Setting

$$\phi(y) = \frac{1}{\sqrt{2}}\sqrt{h(y)}e^{-ig(y)}/2$$

and asking that g, h be analytic in y and real for y real, i.e. $g^*(y) = g(y)$, $h^*(y) = h(y)$, we get Moore's conservation laws [2]:

$$\begin{aligned} \frac{\partial}{\partial t} h &= \frac{\partial}{\partial y} g, \\ \frac{\partial}{\partial t} g &= h^{-2} \frac{\partial}{\partial y} h. \end{aligned} \tag{11}$$

As an example of initial data, suppose that

$$S(\gamma, t = 0) = i\varepsilon \sin \gamma$$

which is Moore's initial data. Then

$$f(y, 0) = \frac{\varepsilon}{2} e^y$$

and therefore:

$$\begin{aligned} \phi(y, 0) &= 1 + \frac{i\varepsilon}{2} e^y \\ \psi(y, 0) &= 1 - \frac{i\varepsilon}{2} e^y \end{aligned}$$

is the initial data for (10) and

$$\begin{aligned} h(y, 0) &= 2 + \frac{\varepsilon^2}{2} e^{2y} \\ g(y, 0) &= -2 \tan^{-1}\left(\frac{\varepsilon}{2} e^y\right) \end{aligned}$$

is the initial data for (11).

Considering the system of conservation laws (10) and its corresponding initial data, we were able to show that the solution developed a singularity after a finite time (using P. Lax's a priory estimates for the time of blow up of a 2 by 2 system of conservation law, see P. Lax [13]).

Finally considering (10) with the error terms that came from H_+D and, using the abstract Cauchy Kowalewsky theorem established by L. Nieremberg and improved by Nishida [15] we were able to establish a long time existence theorem for the solution of the Birkhoff-Rott equation (see Caflisch and Orellana [4]).

Also under the assumption of analyticity, Sulem, Sulem, Bardos and Frisch [10] were able to prove short time existence for solutions of the initial value problem (4)-(5). Borgers [16] and Caflisch and Lowengrub [17] proved global existence for the solution of a *desingularized* version of the initial value problem (4)-(5) as proposed by R. Krasny [6].

The need for the analytic function space setting for the vortex sheet problem, has been demostrated by Caflisch and Orellana [7], who showed that the problem is not well-posed in the Sobolev space H^n for any $n > 3/2$. We constructed exact nonlinear solutions with small initial norm for which $\partial^\alpha S$ become infinite in arbitrary short time for any $\alpha > 1$. Such singular solutions where also constructed by Duchon and Robert [18], and an alternative proof of ill-posedness was given by Ebin [19]. The singular solutions produced in [7] are not believed to be typical (i.e. the type of singularity formation at the initial time is still on open question).

Moore [20] has derived corrections to the Birkhoff-Rott equation, for a thin vortex layer approximating a vortex sheet. Numerical solutions and asymptotic analysis for thin layers have been carried out by Baker and Shelley [21] and Shelley and Baker [22].

There are other important analytic aspects about the Birkhoff-Rott equation and its solutions, like similarity solutions of the form

$$Z(\gamma,t) = t^{\frac{1}{2-p}} Z \ \left(\xi = \gamma t^{\frac{p}{2-p}}\right)$$

which has been analysed by a number of authors.

So far, we learn from the analysis that computation of the evolution of vortex sheet in two-dimensional, incompressible inviscid flow is delicate because of two reasons:

a) Kelvin-Helmholtz instability, and

b) Singularity formation on the interface after finite time, infinite curvature of the interface.

Hence any numerical method must take into account this problem, because numerical roundoff error can excite the physical instability to produce irregular results well before the physically correct singularity formation and roll-up of the vortex sheet.

2. The point vortex method, the vortex blob method and convergence of this methods for vortex sheet

The difficulties mentioned above to compute the evolution of vortex sheets were overcame by R. Krasny [5] and [6] by two methods.

a) A point vortex methods with filtering to eliminate spurious high wavenumber components, and

b) A vortex blob method.

From a formal point of view both methods are similar. The former consists of replacing the continuous curve representing the interface or vortex sheet $Z(\gamma, t)$ at a fixed time by a finite number of point vortex, corresponding to a uniform γ-mesh. This is achieved by discretizing the integral part of Birkhoff-Rott equation by Simpson's rule, and then using a Runge-Kutta 4th order method to solve numerically the system of O.D.E. that was obtained after discretization of the integral part of equation (4).

In the process of iteration there is a filter that at each iteration filters the discrete solution to eliminate spurious high wave number components stabilizing the numerical process in time and space which otherwise will produce irregular results (chaotic motion of the point vortices). Using this numerical method R. Krasny simulated singularity formation for the vortex sheet evolution. Moreover, the jump in tangential velocity developed a cusp at the singular points. Krasny's numerical results are in good agreement with Moore's asymptotic results.

The vortex blob method consists of replacing the continuous curve representing the interface or vortex sheet $Z(\gamma,t)$ at a fixed time by a finite number of circular vortex patches of radius δ corresponding to a uniform γ–mesh. This is achived by doing exactly the same type of discretization described above, but this time applied to the desingularized integro-differential equation

$$\frac{\partial \overline{Z}}{\partial t}(\gamma,t) = \frac{1}{2\pi i}\int_{-\infty}^{\infty} \frac{\overline{Z}(\gamma,t)-\overline{Z}(\gamma',t)d\gamma'}{|Z(\gamma,t)-Z(\gamma',t)|^2+\delta^2} \tag{12}$$

(with no filter). Using this method R. Krasny, was able to simulate roll-up of the vortex sheet, giving numerical evidence of the convergence of the method with respect to both discretization parameters (i.e. for the discretization parameters corresponding to the time and space variable respectively) holding δ equal to const. He also gives numerical evidence of the convergence of the vortex blob method to the vortex sheet computing the solution of equation (12) for several values of the parameter δ going to zero and holding the other two parameters constant.

However, no analytic proof of the convergence had been given until Caflisch and Lowengrub [17] proved convergence of both the point vortex method and the vortex blob method with both spatial and temporal discretization and simulated roundoff error. For the vortex point method they proved convergence in the case of a vortex sheet that is initially a small analytic perturbation of a flat, uniform sheet and the perturbation is chosen periodic for simplicity. They proved convergence for a short time.

The roundoff error e_r must satisfy

$$e_r < \max\left\{e^{-1/h}, e^{-1/\delta}\right\}$$

This condition on the roundoff error is quite strict, but it is consistent with the numerical results of Krasny. *This condition is the numerical interpretation of the requirement of analyticity for vortex sheet solutions.* For the spatial discretization size h the maximum wavenumber is $k_m = 1/h$. Analyticity for a function f is (roughly speaking) equivalent to requiring that $\widehat{f}(k) < \exp(-c|k|)$. This condition can be verified for all k with $|k| \leq k_m$ only if the round off error is sufficiently small; i.e. $e_r < \exp(-c/h)$. First of all Caflisch and Lowengrub prove global existance for analytic solutions of Krasny's desingularized vortex sheet equation (12). This result is established without any reference of closeness to a flat vortex sheet and does not use the Cauchy-Kowalewski Theorem, but it is established for arbitrary analytic initial data.

The abstract Cauchy-Kowalewski Theorem as established by Nierenberg and improved by Nishida and a discretized version of it is the basic tool they used to construct the solutions and to prove convergence of the vortex blob method. In the limit $\delta = 0$ they get a convergence proof for the point vortex method and for a short time interval, certainly less than the critical time. Hence there is no convergence proof of the point vortex method after t_c. For this result it was necessary to set the analysis in an analytic function space because of the use of Cauchy-Kowalewsky theorem, but more important because under this assumption the stability of the point vortex method is mantained, otherwise the problem is ill posed as proved in [7]. Hence one of the main contributions of their paper is the clarification of the meaning of analy-ticity for numerical analysis because of its possible application to many other ill-posed problems.

3. A new physically desingularized vortex sheet equation

In the previous section we mentioned that R. Krasny [6] formulated a desingularized approximation equation of Birkhoff-Rott equation for the evolution of a vortex sheet and he used it to compute roll-up of a vortex sheet after the critical time (i.e. after singularity formation). Krasny's desingularized equation is:

$$\partial_t \overline{Z}(\gamma,t) = K_\delta[Z,\overline{Z}] = \frac{1}{2\pi i}\int_{-\infty}^{\infty} \frac{(\overline{Z}(\gamma,t)-\overline{Z}(\gamma',t)d\gamma'}{|Z(\gamma,t)-Z(\gamma',t)|^2+\delta^2} \tag{13}$$

After the analytic extension of $Z(\gamma,t)$ to the complex γ plane (13) can be rewritten:

$$\partial_t Z^*(\gamma,t) = K_\delta[Z,Z^*] = \frac{1}{2\pi i}\int_{-\infty}^{\infty} \frac{Z^*(\gamma,t)-Z^*(\gamma',t)d\gamma'}{|Z(\gamma,t)-Z(\gamma',t)|^2+\delta^2}, \tag{14}$$

where $Z^*(\gamma,t) = \overline{Z}(\overline{\gamma},t)$.

Even though, from a mathematical point of view, when δ goes to zero (13) approaches the Birkhoff-Rott equation, the desingularization proposed by R. Krasny seems rather arbitrary. Moreover, from a physical point of view, δ does not represent the viscosity because the desingularized equation (13) conserves energy, whereas viscosity would dissipate energy. Thus the effect of δ is dispersive rather dissipative.

Here we give an argument that illustrates that δ does not represent a vortex layer thickness either.

If we apply the localization method illustrated in the first section of this paper to equation (14) and the equation of motion of a vortex layer of small thickness derived by Moore [20], namely:

$$\frac{\partial Z^*}{\partial t}(\gamma,t) = M_\varepsilon[Z,Z^*] = \frac{1}{2\pi i}\fint \frac{d\gamma'}{Z(\gamma,t)-Z(\gamma',t)} + \frac{\varepsilon}{6\omega i}\frac{\partial}{\partial\gamma}\left(\left|\frac{\partial Z}{\partial \gamma}\right|^{-4}\frac{\partial Z^*}{\partial\gamma}\right)$$

where the terms of size $O(\varepsilon^2)$ have been ignored; $\varepsilon = H/\rho$, H =the dimensional thickness and ρ the radius of curvature, and $\omega := \varepsilon\overline{\omega}$ is the vortex sheet strength with $\overline{\omega}$ the constant vorticity in the layer; we get:

$$\begin{aligned}
\partial_t S_-^* &= M_\delta[S_-,S_+^*] = \frac{1}{2}\left\{\frac{\partial_\gamma S_+}{1+S_{+\gamma}}\right\} \\
&+ \frac{i\delta}{8}\left\{\frac{\partial_\gamma^2 S_-^*}{(1+\partial_\gamma S_+)^{3/2}(1+\partial_\gamma S_-^*)^{3/2}} + \frac{3\partial_\gamma^2 S_+}{(1+\partial_\gamma S_+)^{5/2}(1+\partial_\gamma S_-^*)^{1/2}}\right\}, \\
\partial_t S_+^* &= M_\delta[S_-,S_+^*]^* = -\frac{1}{2}\left\{\frac{\partial_\gamma S_-^*}{1+\partial_\gamma S_-^*}\right\} \\
&- \frac{i\delta}{8}\left\{\frac{\partial_\gamma^2 S_+}{(1+\partial_\gamma S_-^*)^{3/2}(1+\partial S_+)^{3/2}} + \frac{3\partial_\gamma^2 S_-}{(1+\partial_\gamma S_-^*)^{5/2}(1+\partial_\gamma S_+)^{1/2}}\right\},
\end{aligned}$$

and

$$
\begin{aligned}
\partial_t S^*_- &= M_\varepsilon[S_+, S^*_-] \\
&= \frac{1}{2}\left\{\frac{\partial_\gamma S_+}{1+\partial_\gamma S_+}\right\} + \varepsilon(6\omega i)^{-1}\partial_\gamma\left\{(1+\partial_\gamma S^*_+)^{-2}(1+\partial_\gamma S^*_-)^{-1}\right\} \\
\partial_t S_+ &= M_\varepsilon[S_+, S^*_-]^* \\
&= -\frac{1}{2}\left\{\frac{\partial_\gamma S^*_-}{1+\partial_\gamma S^*_-}\right\} - \varepsilon(6\omega i)^{-1}\partial_\gamma\left\{(1+\partial_\gamma S^*_-)^{-2}(1+\partial_\gamma S_+)^{-1}\right\}
\end{aligned}
$$

respectively.

Comparison of these equations shows them to be quite different, hence the desingularization parameter δ cannot be interpreted as vortex layer thickness. This difference can be understood by noting that Moore's expansion parameter is ε/ω which has units of (length/velocity).

A more physically meaningful desingularization equation is found by replacing δ^2 in (13) or (14) by $(\varepsilon/\omega)^2$. Then a factor with units of velocity must be put into the first term of the denominator. Hence we proposed the following physically desingularized vortex sheet equation:

$$
\partial_t \overline{Z}(\gamma,t) = \frac{1}{2\pi i}\int_{-\infty}^{\infty} \frac{\left(\overline{Z}(\gamma,t)-\overline{Z}(\gamma',t)\right)\left(\left|\frac{\partial Z}{\partial\gamma}(\gamma,t)\right|^2+\left|\frac{\partial Z}{\partial\gamma}(\gamma',t)\right|^2\right)d\gamma'}{|Z(\gamma,t)-Z(\gamma',t)|^2\left\{\left|\frac{\partial Z}{\partial\gamma}(\gamma,t)\right|^2+\left|\frac{\partial Z}{\partial\gamma}(\gamma',t)\right|^2\right\}+\frac{8}{9}\left(\frac{\varepsilon}{\omega}\right)^2}
$$

which agrees up to size $O(\frac{\varepsilon}{\omega})^2$ with the vortex layer equation in the localized approximate equation.

Analytic and numerical analysis of this last equation is presently being performed.

Conclusion

From what has been presented here the importance of taking into account all the aspects involved in the resolution of a given problem is quite obvious.

In particular, for this problem, it is useful to notice how the analytic, numerical and physical aspects of the problem play an important role in its resolution and how they relate and complement one other to validate the different solutions and finally the model in question (i.e. Birkhoff-Rott equation as a mathematical model to describe the evolution and dynamics of a vortex sheet).

References

[1] D.W. Moore, *The spontaneous appearance of a singularity in the shape of an evolving vortex sheet*, Proc. Roy. Soc. London, Ser. A, **365** (1979), 105-119.

[2] D.W. Moore, *Numerical and Analytical aspects of Helmholtz instability in Theoretical and Applied Mechanics* Proc. XVI Internat. Congr. Theoret. Appl. Mech., F.I. Niordson and N. Olhoff, eds., North-Holland, Amsterdam, 1984, 629-633.

[3] D.I. Meiron, R.G. Baker, S.A. Orszag, *Analytic structure of vortex sheet dynamics*, Part 1, Kelvin-Helmholtz instability, J. Fluid Mech. **114** (1982) 283-298.

[4] R.E. Caflisch, O. Orellana. *Long Time Existence for a Slightly Perturbed Vortex Sheet*, Comm. Pure Appl. Math. **39** (1986) 807-838.

[5] R. Krasny, *On singularity formation in a vortex sheet and the point vortex approximation*, J. Fluid Mech. **167**, (1986) 65-93.

[6] Krasny, R., Desingularization of periodic vortex sheet roll-up, J. Comp. Phys. **65**, (1986) 292-313.

[7] R.E. Caflisch, O. Orellana, *Singular Solutions and Ill-Posedness for the Evolution of Vortex Sheets*, SIAM J. Math. Anal. **20**, (1989) 293-307.

[8] H.W. Hoeigmakers, W. Vaatstra, W., *A higher order panel method applied to vortex sheet roll-up*, J. AIAA **21**, (1983) 516–523.

[9] G. Birkhoff, *Helmholtz and Taylor instability* in "Hydrodynamic Instability", Proc. Symp. in Appl. Math. XII, AMS (1962), 55–76

[10] C. Sulem, P.L. Sulem, C. Bardos, U. Frisch, *Finite time analyticity for the two and three dimensional Kelvin–Helmholtz instability*, Comm. Math. Phys. **80**, (1981) 485-516.

[11] G.K. Batchelor, *An introduction to Fluid Dynamics*, Cambridge University Press (1967) 511-517.

[12] P. Garabedian, *Partial differential equations*, John Wiley and Sons, 1964.

[13] P.D. Lax, *Development of singularities of solutions of nonlinear hyperbolic partial differential equations*, J. Math. Phys. **5**, (1964), 611-613.

[14] R. E. Caflisch, O. Orellana, M. Siegel, *A Localized Approximation Method for Vortical Flows*, SIAM Journal on Applied Mathematics, (to appear).

[15] T. Nishida, *A note on a theorem of Niremberg*, J. Diff. Geom. **12** (1977) 629-633.

[16] C. Borges, *On the numerical solution of the regularized Birkhoff equation*, preprint 1988.

[17] R.E. Caflisch, J. Lowengrub, *Convergence of the vortex method for vortex sheets*, SIAM J. Num., Anal. (to appear).

[18] J. Duchon, R. Robert, *Solutions globales avec nape tourbillionaire pour les equations d'Euler dans le plan*, C.R. Acad. Sci., Paris **302** (1986) 183–186.

[19] D. Ebin, *Ill-posedness of the Rayleigh-Taylor and Helmholtz problems for incompressible fluids*, Comm. P.D.E. **13** (1985) 1265-1295.

[20] D.W. Moore, *The equation of motion of a vortex layer of small thickness*, Stud. in Appl. Math. **58**, (1978) 119-140.

[21] G.R. Baker, M.J. Shelley, *On the connection between thin vortex layers and vortex sheets*, Part I: J. Fluid Mech. (to appear).

[22] M.J. Shelley, Baker, G.R., *On the connection between thin vortex layer and vortex sheets*, Part II: Numerical Study, J. Fluid Mech. (to appear).

Received: July 30, 1989

Computational Methods and Function Theory
Proceedings, Valparaíso 1989
St. Ruscheweyh, E.B. Saff, L. C. Salinas, R.S. Varga (*eds.*)
Lecture Notes in Mathematics **1435**, pp. 155–169

On the Numerical Performance of a Domain Decomposition Method for Conformal Mapping

N. Papamichael and N.S. Stylianopoulos

Department of Mathematics and Statistics, Brunel University
Uxbridge, Middlesex UB8 3PH, U.K.

1. Introduction

This paper is a sequel to a recent paper [14], concerning a domain decomposition method (hereafter referred to as *DDM*) for the conformal mapping of a certain class of quadrilaterals. For the description of the *DDM* we proceed exactly as in [14:§1], by introducing the following terminology and notations.

Let G be a simply-connected Jordan domain in the complex z-plane ($z = x+iy$), and consider a system consisting of G and four distinct points z_1, z_2, z_3, z_4 in counterclockwise order on its boundary ∂G. Such a system is said to be a quadrilateral Q and is denoted by $Q = \{G; z_1, z_2, z_3, z_4\}$. The conformal module $m(Q)$ of Q is defined as follows:

Let R be a rectangle of the form

$$R := \{(\xi, \eta) : a < \xi < b, c < \eta < d\}, \tag{1.1}$$

in the w-plane ($w = \xi + i\eta$), and let h denote its aspect ratio, i.e. $h := (d-c)/(b-a)$. Then $m(Q)$ is the unique value of h for which Q is conformally equivalent to a rectangle of the form (1.1), in the sense that for $h = m(Q)$ and for this value only there exists a unique conformal map $R \to G$ which takes the four corners $a+ic$, $b+ic$, $b+id$, and $a+id$, of R respectively onto the four points z_1, z_2, z_3, z_4. In particular, $h = m(Q)$ is the only value of h for which Q is conformally equivalent to a rectangle of the form

$$R_h\{\alpha\} := \{(\xi, \eta) : 0 < \xi < 1, \alpha < \eta < \alpha + h\}. \tag{1.2}$$

The *DDM* is a method for computing approximations to the conformal modules and associated conformal maps of quadrilaterals of the form illustrated in Figure 1.1(b). That is, the method is concerned with the mapping of quadrilaterals

$$Q := \{G; z_1, z_2, z_3, z_4\}, \tag{1.3a}$$

where:

• The domain G is bounded by the straight lines $x = 0$ and $x = 1$ and two Jordan arcs with cartesian equations $y = -\tau_1(x)$ and $y = \tau_2(x)$, where τ_j; $j = 1, 2$, are positive in $[0, 1]$, i.e.

$$G := \{(x, y) : 0 < x < 1, -\tau_1(x) < y < \tau_2(x)\}. \tag{1.3b}$$

• The points z_1, z_2, z_3, z_4 are the corners where the arcs intersect the straight lines, i.e.

$$z_1 = -i\tau_1(0), \quad z_2 = 1 - i\tau_1(1), \quad z_3 = 1 + i\tau_2(1), \quad z_4 = i\tau_2(0). \tag{1.3c}$$

Let Q be of the form (1.3) and let

$$G_1 := \{(x, y) : 0 < x < 1, -\tau_1(x) < y < 0\}, \tag{1.4a}$$

and

$$G_2 := \{(x, y) : 0 < x < 1, 0 < y < \tau_2(x)\}, \tag{1.4b}$$

so that $\overline{G} = \overline{G}_1 \cup \overline{G}_2$. Also, let Q_1 and Q_2 denote the quadrilaterals

$$Q_1 := \{G_1; z_1, z_2, 1, 0\} \quad \text{and} \quad Q_2 := \{G_2; 0, 1, z_3, z_4\}, \tag{1.4c}$$

and let $h := m(Q)$ and $h_j := m(Q_j)$; $j = 1, 2$; see Figures 1.2(b) and 1.3(b). Finally, let g and g_j; $j = 1, 2$, denote the conformal maps

$$g : R_h\{-h_1\} \to G, \tag{1.5}$$

$$g_1 : R_{h_1}\{-h_1\} \to G_1 \quad \text{and} \quad g_2 : R_{h_2}\{0\} \to G_2, \tag{1.6}$$

where, with the notation (1.2),

$$R_h\{-h_1\} := \{(\xi, \eta) : 0 < \xi < 1, -h_1 < \eta < h - h_1\},$$

$$R_{h_1}\{-h_1\} := \{(\xi, \eta) : 0 < \xi < 1, -h_1 < \eta < 0\},$$

$$R_{h_2}\{0\} := \{(\xi, \eta) : 0 < \xi < 1, 0 < \eta < h_2\};$$

see Figures 1.1-1.3. Then, the *DDM* consists of the following:

(a) Subdividing the quadrilateral Q, given by (1.3), into the two smaller quadrilaterals Q_1 and Q_2, given by (1.4).

(b) Approximating the conformal module of Q by the sum of the conformal modules of Q_1 and Q_2, i.e. approximating h by

$$\tilde{h} := h_1 + h_2. \tag{1.7a}$$

(c) Approximating the rectangle $R_h\{-h_1\}$ and the conformal map $g : R_h\{-h_1\} \to G$ respectively by

$$R_{\tilde{h}}\{-h_1\} := \{(\xi,\eta) : 0 < \xi < 1, -h_1 < \eta < h_2\}, \tag{1.7b}$$

and

$$\tilde{g}(w) := \begin{cases} g_2(w) : R_{h_2}\{0\} \to G_2, & \text{for } w \in R_{h_2}\{0\}, \\ g_1(w) : R_{h_1}\{-h_1\} \to G_1, & \text{for } w \in R_{h_1}\{-h_1\}. \end{cases} \tag{1.7c}$$

The initial motivation for considering the above method came from: (a) The intuitive observation that if the constituent quadrilaterals Q_1 and Q_2 are both "long" then $\tilde{h}$ is close to h. (b) Experimental evidence indicating that $\tilde{h}$ is close to h even when Q_1 and Q_2 are only moderately long; see [12:§5] and [14:§1]. (It is important to note that $h \geq h_1 + h_2$ and equality occurs only in the trivial cases where G is a rectangle or $\tau_1(x) = \tau_2(x)$, $x \in [0,1]$; see e.g. [9: p. 437].)

The treatment of the *DDM* contained in [14] is a theoretical investigation leading to estimates of the errors in the approximations (1.7). These error estimates are derived by assuming that the functions τ_j; $j = 1,2$, satisfy the following:

(i) τ_j; $j = 1,2$, are absolutely continuous in [0,1], and

$$d_j := \operatorname*{ess\,sup}_{0\leq x\leq 1} |\tau_j'(x)| < \infty. \tag{1.8}$$

(ii) If

$$m_j := \max_{0\leq x\leq 1}\{\exp(-\pi\tau_j(x))\}; \quad j := 1,2, \tag{1.9a}$$

then

$$\varepsilon_j := d_j\{(1+m_j)/(1-m_j)\} < 1; \; j = 1,2. \tag{1.9b}$$

In addition to the theory, [14] also contains two numerical examples comparing the actual errors in the *DDM* approximations with those predicted by the theoretical estimates.

The present paper is concerned with the numerical performance of the *DDM* and, in particular, with the performance of the method in cases where the functions τ_j; $j = 1,2$, do not fulfil the rather restrictive condition (1.9) needed for the theory of [14]. More specifically, the main purpose of this paper is to show by means of numerical examples that some of the theoretical results of [14] remain valid even when the condition (1.9) is violated, and thus to provide experimental support for certain conjectures made in [14].

We end this introductory section by making the following remarks concerning the *DDM* and related matters:

- A survey of available methods for computing approximations to the conformal modules and the associated conformal maps of general quadrilaterals is given in [12], where also several areas of application of the conformal maps are discussed; see also [4]-[7] and [9:§16.11].
- Although this is not considered here, the *DDM* can also be applied to quadrilaterals of the form illustrated in Figure 1.4, provided that the crosscut c of subdivision is taken

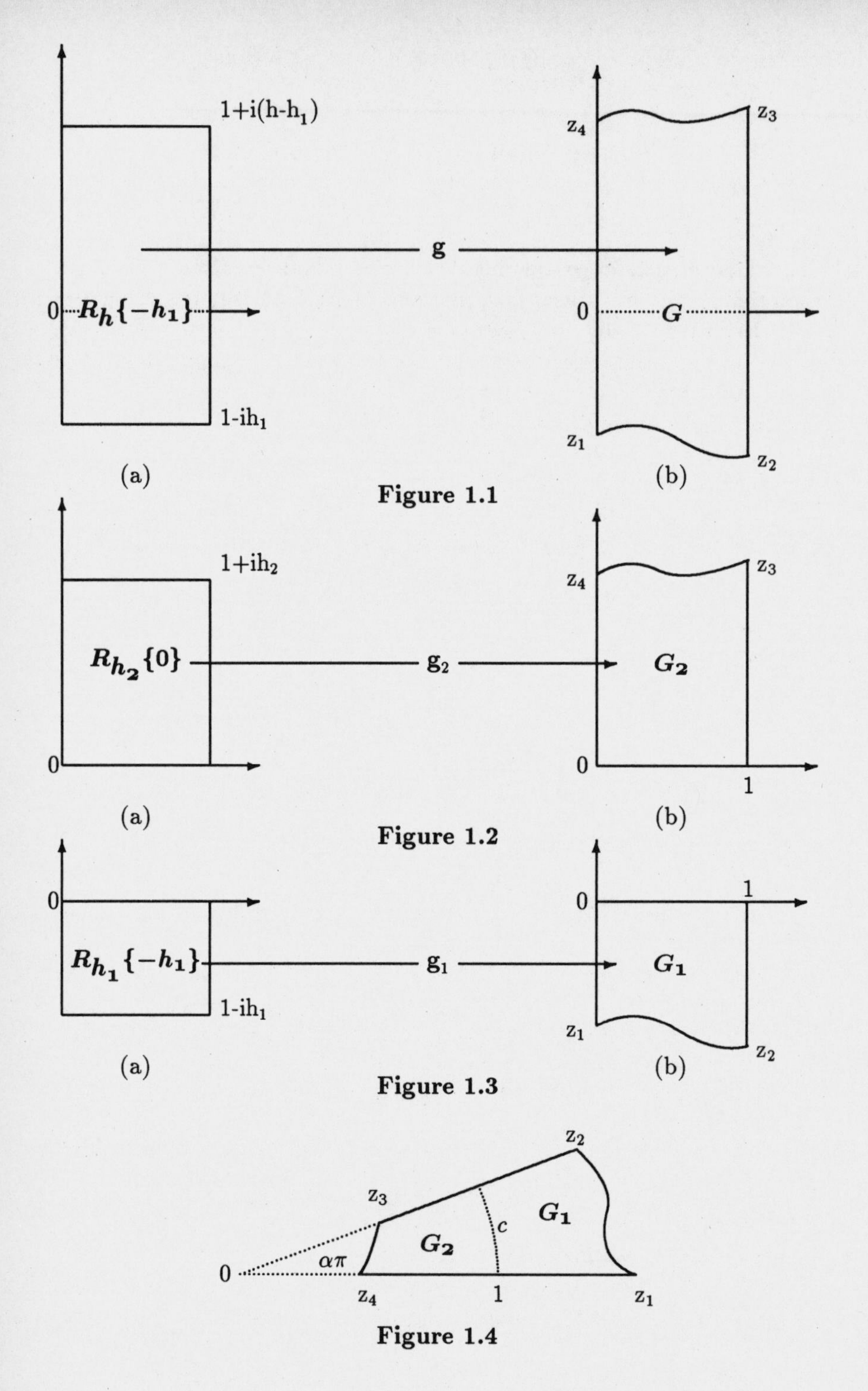

Figure 1.1

Figure 1.2

Figure 1.3

Figure 1.4

to be a circular arc; see [14: Remark 4.7]. In other words, the application of the *DDM* is restricted to quadrilaterals that have one of the two special forms illustrated in Figures 1.1 and 1.4. We note however that the mapping of such quadrilaterals has received considerable attention recently; see e.g. [2], [8], [11], [12], [15] and [17].

• A more general form of the *DDM* involves subdividing the original quadrilateral Q into two quadrilaterals Q_1 and Q_2 of the form (1.4) at the lower and upper ends, and a rectangle in the middle. This can be described more precisely as follows:

Let

$$G := \{(x,y) : 0 < x < 1, -\tau_1(x) < y < \tau_2(x) + c\},$$

where $c > 0$, let

$$G_1 := \{(x,y) : 0 < x < 1, -\tau_1(x) < y < 0\},$$

and

$$G_2 := \{(x,y) : 0 < x < 1, c < y < \tau_2(x) + c\},$$

so that $\overline{G} = \overline{G_1} \cup \overline{R_c}\{0\} \cup \overline{G_2}$, and let

$$z_1 = -i\tau_1(0),\ z_2 = 1 - i\tau_1(1),\ z_3 = 1 + i(\tau_2(1) + c),\ z_4 = i(\tau_2(0) + c).$$

Then the general form of the *DDM* consists of the following:

(a) Subdividing the quadrilateral $Q := \{G; z_1, z_2, z_3, z_4\}$ into three smaller quadrilaterals, i.e. the quadrilaterals

$$Q_1 := \{G_1; z_1, z_2, 1, 0\} \quad \text{and} \quad Q_2 := \{G_2; ic, 1 + ic, z_3, z_4\},$$

at the lower and top ends, and the rectangular quadrilateral

$$\{R_c\{0\}; 0, 1, 1 + ic, ic\},$$

in the middle.

(b) Approximating the conformal module $h := m(Q)$ by

$$\tilde{h} := h_1 + h_2 + c, \tag{1.10}$$

where $h_j := m(Q_j)$; $j = 1, 2$.

(c) Approximating the rectangle $R_h\{-h_1\}$ and the conformal map $g : R_h\{-h_1\} \to G$ respectively by $R_{\tilde{h}}\{-h_1\}$ and

$$\tilde{g}(w) := \begin{cases} g_2(w) : R_{h_2}\{c\} \to G_2, & \text{for } w \in R_{h_2}\{c\}, \\ w, & \text{for } w \in R_c\{0\}, \\ g_1(w) : R_{h_1}\{-h_1\} \to G_1, & \text{for } w \in R_{h_1}\{-h_1\}. \end{cases} \tag{1.11}$$

• The *DDM* is of practical interest for the following reasons:

(i) Given the conformal modules and associated conformal maps of two quadrilaterals Q_1 and Q_2 of the form (1.4), the method provides approximations to the conformal module and associated conformal map of any quadrilateral consisiting of Q_1 and Q_2, at the lower and top ends, and a rectangle of any height in the middle.

(ii) The method can be used to overcome the "crowding" difficulties associated with the numerical conformal mapping of "long" quadrilaterals of the form (1.3). (Full details of the crowding phenomenon and its damaging effects on numerical procedures for the mapping of "long" quadrilaterals can be found in [12], [13], and [10]; see also [4:p.179], [9:p.428] and [16:p.4].)

(iii) Numerical methods for approximating the conformal maps of quadrilaterals of the form (1.4) are often substantially simpler than those for quadrilaterals of the more general form (1.3); see e.g. [8], and [12,§3.4].

• Of the two conditions involved in the assumptions (1.8)-(1.9), only (1.9) is restrictive from the parctical point of view. This condition is more or less equivalent to requiring that the slopes of the two curves $y = \tau_j(x)$; $j = 1,2$, are numerically less than unity in [0,1]. This is so because the values $m_j; j = 1,2$, given by (1.9a) are "small", even when the quadrilaterals Q_j; $j = 1,2$, are only moderately "long".

2. Theoretical error estimates

As in Section 1, let Q and Q_j; $j = 1,2$, denote the three quadrilaterals defined by (1.3) and (1.4), let $h := m(Q)$, $h_j := m(Q_j)$; $j = 1,2$, and let g, g_j; $j = 1,2$, be the associated conformal maps (1.5) and (1.6). Also, let

$$X(\xi) := \operatorname{Re} g(\xi - ih_1), \quad \hat{X}(\xi) := \operatorname{Re} g(\xi + i(h - h_1)),$$

and

$$X_1(\xi) := \operatorname{Re} g_1(\xi - ih_1), \quad \hat{X}_2(\xi) := \operatorname{Re} g_2(\xi + ih_2),$$

and let E_h and $E_g\{j\}$, $E_X\{j\}$, $E_\tau\{j\}$; $j = 1,2$, denote the following domain decomposition errors:

$$E_h := h - (h_1 + h_2), \tag{2.1}$$

$$E_g\{1\} := \max\{|g(w) - g_1(w)| : w \in \overline{R_{h_1}}\{-h_1\}\}, \tag{2.2a}$$

$$E_g\{2\} := \max\{|g(w + iE_h) - g_2(w)| : w \in \overline{R_{h_2}}\{0\}\}, \tag{2.2b}$$

$$\begin{aligned} E_X\{1\} &:= \max_{0\le\xi\le1} |X(\xi) - X_1(\xi)|, \\ E_X\{2\} &:= \max_{0\le\xi\le1} |\hat{X}(\xi) - \hat{X}_2(\xi)|, \end{aligned} \tag{2.3}$$

and

$$E_\tau\{1\} := \max_{0\le\xi\le1} |\tau_1(X(\xi)) - \tau_1(X_1(\xi))|, \quad E_\tau\{2\} := \max_{0\le\xi\le1} |\tau_2(\hat{X}(\xi)) - \tau_2(\hat{X}_2(\xi))|. \tag{2.4}$$

Finally, assume that the functions τ_j; $j = 1,2$, satisfy the assumptions (1.8)-(1.9), and let

(2.5a) $$\alpha(\varepsilon_j,\varepsilon) := 2/\{(1-\varepsilon_j)(1-\varepsilon^2)^{\frac{1}{2}}\};\ j=1,2,$$

and

(2.5b) $$\beta(\varepsilon_j,\varepsilon) := \sqrt{8}/\{(1-\varepsilon_j)(1-\varepsilon^2)\}^{\frac{1}{2}};\ j=1,2,$$

where the ε_j; $j=1,2$, are given by (1.9), and $\varepsilon := \max(\varepsilon_1,\varepsilon_2)$. Then, the main results of [14] are the following estimates of the errors E_h and $E_g\{j\}$; $j=1,2$, in the *DDM* approximations (1.7):

(2.6) $$E_h \leq \pi^{-1}d_1\alpha(\varepsilon_1,\varepsilon)\{\varepsilon_1 e^{-2\pi h_1}+\varepsilon_2 e^{-\pi h}\}+\pi^{-1}d_2\alpha(\varepsilon_2,\varepsilon)\{\varepsilon_2 e^{-2\pi h_2}+\varepsilon_1 e^{-\pi h}\}$$

and

(2.7a) $$E_g\{j\} \leq \max\{M_j,N_j\};\ j=1,2,$$

where

(2.7b) $$M_j := \pi^{-\frac{1}{2}}(1+d_j^2)^{\frac{1}{2}}\beta(\varepsilon_j,\varepsilon)\{\varepsilon_j+\varepsilon_{3-j}e^{-\pi h}\}^{\frac{1}{2}}\{\varepsilon_j e^{-2\pi h_j}+\varepsilon_{3-j}e^{-\pi h}\}^{\frac{1}{2}},$$

(2.7c) $$\begin{aligned} N_j &:= \tfrac{1}{2}\pi^{-\frac{1}{2}}\beta(0,\varepsilon)\{5\varepsilon_j e^{-\pi h_j}+3\varepsilon_{3-j}e^{-\pi(h-h_j)}\}\\ &\quad +\pi^{-1}d_j\alpha(\varepsilon_j,\varepsilon)\{\varepsilon_j e^{-2\pi h_j}+\varepsilon_{3-j}e^{-\pi h}\};\end{aligned}$$

see [14:Thms 4.1, 4.4]. Since $h \geq h_1+h_2$, the above estimates show that if the functions τ_j; $j=1,2$, satisfy the assumptions (1.8)-(1.9), then

(2.8) $$E_h = O\{\exp(-2\pi h^*)\},$$

and

(2.9) $$E_g\{j\} = O\{\exp(-\pi h^*)\};\quad j=1,2,$$

where $h^* := \min(h_1,h_2)$.

In addition to (2.6) and (2.7), [14] also contains estimates of the errors $E_X\{j\}$ and $E_\tau\{j\}$; $j=1,2$; see Theorem 4.2 and Remark 4.2 in [14]. These estimates show that under the assumptions (1.8)-(1.9),

(2.10) $$E_X\{j\} = O\{\exp(-\pi h^*)\}\quad\text{and}\quad E_\tau\{j\} = O\{\exp(-\pi h^*)\};\quad j=1,2,$$

where as before $h^* := \min(h_1,h_2)$. Finally, [14] contains a theorem (Theorem 4.3), which shows that if Q_1 and Q_2 are both "long" quadrilaterals, then at points sufficiently far from the two sides $\eta=-h_1$ and $\eta=h-h_1$ of $R_h\{-h_1\}$, the conformal map g can be approximated closely by the identity map. In particular, this theorem of [14] shows that

(2.11) $$\max_{0\leq\xi\leq 1}|g(\xi+i0)-\xi| = O\{\exp(-\pi h^*)\}.$$

The method of analysis used in [14] for deriving the above results makes extensive use of the theory given in [3: Kap.V,§3], in connection with the integral equation method of Garrick, for the confomal mapping of doubly-connected domains. This involves expressing the three problems for the conformal maps g and g_j; $j = 1,2$, as equivalent problems for the conformal maps of three symmetric doubly-connected domains; see [8], [12:§3.2,3.4] and [13:§3]. (With reference to the above comment, readers who are familiar with the method of Garrick will recognize the very close similarity between the condition (1.9), which is needed for the analysis used in [14], and the so-called $\varepsilon\delta$-condition needed for the theory of the Garrick method.)

We end this section by observing that the results of [14] simplify considerably in the case where one of the two subdomains G_1 of G_2 is a rectangle. For example, let $\tau_1(x) = c > 0$, $x \in [0,1]$, i.e. let

$$G_1 := \{(x,y) : 0 < x < 1, -c < y < 0\} = R_c\{-c\}. \tag{2.12}$$

Then, $g_1(w) = w$, $h_1 = c$, $d_1 = \varepsilon_1 = 0$ and, for any value $c > 0$, the results (2.8)-(2.10) simplify as follows:

$$E_h := h - (c + h_2) = O\{\exp(-2\pi h_2)\}, \tag{2.13}$$

$$E_g\{j\} = O\{\exp(-\pi h_2)\}; \quad j = 1,2, \tag{2.14}$$

$$E_X\{j\} = O\{\exp(-\pi h_2)\}, \quad E_\tau\{j\} = O\{\exp(-\pi h_2)\}; \quad j = 1,2. \tag{2.15}$$

Also, in place of (2.11) we now have for any point $w := \xi + i\eta \in R_c\{-c\}$,

$$|g(w) - w| = O\{\exp(-\pi(h_2 - \eta))\}. \tag{2.16}$$

Furthermore, all the above simplified results hold under the less restrictive assumptions obtained by replacing the inequalities (1.9) by

$$\varepsilon_2 := d_2\{(1 + m_2^2)/(1 - m_2^2)\} < 1; \tag{2.17}$$

see [14:Remark 4.5].

3. Numerical results and discussion

In addition to the theoretical estimates summarized in Section 2, [14:§5] contains two numerical examples in which the quadrilaterals are chosen so that the functions τ_j; $j = 1,2$, satisfy the assumptions (1.8)-(1.9). The numerical results of these examples confirm the theory of [14], and indicate that the *DDM* is capable of producing approximations of high accuracy, even when the quadrilaterals under consideration are only moderately long.

In this section we study further the numerical performance of the *DDM*, but here we consider its application to quadrilaterals that do not satisfy the conditions (1.9). That is, we are concerned with cases for which the theory of [14] does not apply. Our main purpose is to provide experimental evidence supporting the following two conjectures made in [14]:

Conjecture 3.1. *The results (2.8)-(2.9) hold even when the condition (1.9) is not fulfilled; see [14:Remark 5.4]. More specifically, the claim here is that*

$$E_h = O\{\exp(-2\pi h^*)\}, \tag{3.1}$$

and

$$E_g\{j\} = O\{\exp(-\pi h^*)\}; \quad j = 1, 2, \tag{3.2}$$

with $h^* := \min(h_1, h_2)$ *even when* $d_j \geq 1$; $j = 1, 2$, *where* d_j *are the values given by (1.8).*

Conjecture 3.2. *The errors* $E_X\{j\}$ *and* $E_\tau\{j\}$; $j = 1, 2$, *are* $O\{\exp(-2\pi h^*)\}$, *rather than* $O\{\exp(-\pi h^*)\}$ *as predicted by the theory of [14]; see (2.10) and [14:Remark 5.2]. That is, the claim here is that*

$$E_X\{j\} = O\{\exp(-2\pi h^*)\} \quad \text{and} \quad E_\tau\{j\} = O\{\exp(-2\pi h^*)\}; \quad j = 1, 2, \tag{3.3}$$

and that the above results hold even when the condition (1.9) is violated.

Each of the three examples considered below involves the mapping of a quadrilateral Q of the form (1.3) and, in each case, the decomposition is performed by subdividing Q into two quadrilaterals Q_j; $j = 1, 2$, of the form (1.4). In presenting the numerical results we use the following notations:

- E_h and $E_g\{j\}$, $E_X\{j\}$, $E_\tau\{j\}$; $j = 1, 2$: As before these denote the actual *DDM* errors (2.1)-(2.4). More precisely, the values listed in the examples are reliable estimates of the errors (2.1)-(2.4). They are determined, as in [14:§5], from accurate approximations to h, h_j, $j = 1, 2$, and g, g_j, $j = 1, 2$, which are computed by using the iterative algorithms of [8]. In particular, $E_g\{j\}$; $j = 1, 2$, are the maxima of two sets of values which are obtained by sampling respectively the approximations to the functions $g(w) - g_1(w)$ and $g(w + iE_h) - g_2(w)$ at a number of test points on the boundary segments $\eta = -h_1, 0$ of $R_{h_1}\{-h_1\}$ and $\eta = 0, h_2$ of $R_{h_2}\{0\}$. The values $E_X\{j\}$ and $E_\tau\{j\}$ are determined in a similar manner, by sampling the approximations to the functions $X(\xi) - X_1(\xi)$, $\hat{X}(\xi) - \hat{X}_2(\xi)$, etc. at a number of test points in $0 \leq \xi \leq 1$. (The only exception to the above are the values of E_h given in Example 3.1, in which Q is a trapezium and the subdivision consists of a smaller trapezium Q_2 and a rectangle $Q_1 := R_c\{-c\}$. In this case, $h := m(Q)$ and $h_2 := m(Q_2)$ are known exactly in terms of elliptic integrals. Hence $E_h := h - (c + h_2)$ is also known exactly.)
- δ_h and $\delta_g\{j\}$, $\delta_X\{j\}$, $\delta_\tau\{j\}$; $j = 1, 2$: These denote the values used for testing the validity of (3.1)-(3.3). They are determined from the computed values of the errors E_h and $E_g\{j\}$, $E_X\{j\}$, $E_\tau\{j\}$; $j = 1, 2$, as follows:

In each of the Examples 3.2 and 3.3, the functions τ_j; $j = 1, 2$, are of the form

$$\tau_j(x) := \sigma_j(x) + l; \quad j = 1, 2,$$

where $l \geq 0$, and in each case the values of the conformal modules and the errors in the *DDM* approximations are computed for several values of the parameter l. Let $h_1(l)$ and $h_2(l)$ denote the conformal modules of Q_1 and Q_2 corresponding to the value l, and let $h^*(l) := \min(h_1(l), h_2(l))$. Also, let E denote any of the errors E_h, $E_g\{j\}$, $E_X\{j\}$ or $E_\tau\{j\}$, and let $E(l)$ be the value of E corresponding to l. Then, the validity of (3.1)-(3.3) is checked by assuming that

$$E = O\{\exp(-\delta\pi h^*)\}; \quad h^* := \min(h_1, h_2),$$

and computing various values of δ (i.e. of δ_h, $\delta_g\{j\}$, $\delta_X\{j\}$ or $\delta_\tau\{j\}$) by means of the formula

$$\delta = -\{\log[E(l_1)/E(l_2)]\}/\{\pi[h^*(l_1) - h^*(l_2)]\}.$$

(In the examples, l_1 and l_2 are taken to be successive values of the paremeter l for which numerical results are listed.)

In the first example, i.e. in Example 3.1, we consider only the error E_h and, because of the form of Q, we check the validity of (2.13) rather than (3.1). That is we assume that

$$E_h = O\{\exp(-\delta_h \pi h_2)\},$$

and determine δ_h from the listed values of E_h, by modifying in an obvious manner the procedure described above.

Example 3.1 (See also [12:§5]) Q is the trapezium illustrated in Figure 3.1. That is, Q is defined by the functions

$$\tau_1(x) = c \quad \text{and} \quad \tau_2(x) = x - 1 + l,$$

where $c > 0$ and $l > 1$. Here, $d_2 = 1$ and, because of this, the theory of [14] does not apply.

As was previously remarked, in this case the conformal modules $h := m(Q)$ and $h_2 := m(Q_2)$ are known in terms of elliptic integrals. Thus, Table 3.1 contains the exact values of h and h_2 corresponding to the parameters l =1.25, 2.00, 2.50, 4.00, 5.00, and c =0.75, 0.50, 1.50, 1.00, 5.00. (These were determined correct to twelve decimal places, by using the formulae of Bowman [1:p.104].) The table also contains the values of the error $E_h := h-(h_2+c)$ and, where possible, the corresponding values of δ_h. These values of δ_h indicate clearly that, for any $c > 0$, $E_h = O\{\exp(-2\pi h_2)\}$.

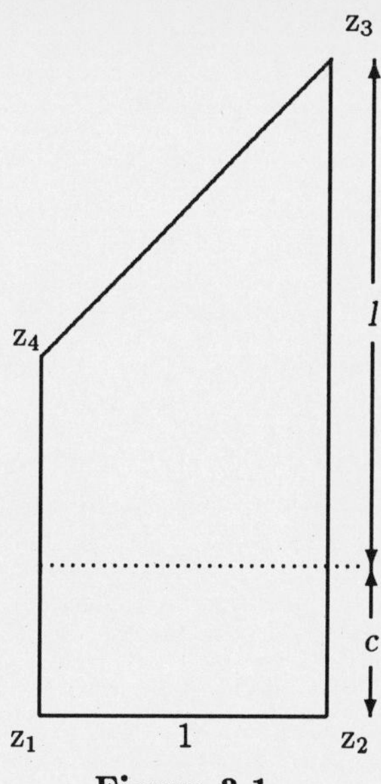

Figure 3.1

l	c	h_2	h	E_h	δ_h
1.25	0.75	0.516 810 878 029	1.297 261 571 171	1.2 E-02	-
2.00	0.50	1.279 261 571 171	1.779 359 959 478	9.8 E-05	2.021
2.50	1.50	1.779 359 959 478	3.279 364 399 489	4.4 E-06	1.972
4.00	1.00	3.279 364 399 489	4.279 364 399 847	3.6 E-10	1.999
5.00	5.00	4.279 364 399 847	9.279 364 399 847	-	-

Table 3.1

Example 3.2. Q and Q_j; $j = 1,2$, are defined by (1.3) and (1.4) with

$$\tau_1(x) = 1 + 0.25\cos(2\pi x) + l \quad \text{and} \quad \tau_2(x) = 0.25x^4 - 0.5x^2 + 1 + l.$$

Since $d_1 = \pi/2 > 1$, the above two functions do not satisfy the condition (1.9) needed for the theory of [14].

The numerical results corresponding to the values $l = 0.0(0.5)2.5$ are listed in Tables 3.2(a)-3.2(c). Table 3.2(a) contains the computed values of the conformal modules, together with the estimates of $E_h := h - (h_1 + h_2)$ and the corresponding values of δ_h. (The values of h and h_j listed in the table are expected to be correct to all the figures quoted. The algorithms of [8] achieve this remarkable accuracy, because the two curves $y = \tau_j(x)$; $j = 1,2$, intersect the straight lines $x = 0$ and $x = 1$ at right angles; see [8:§6]) Tables 3.2(b) and 3.2(c) contain respectively the estimates of the errors $E_g\{j\}$; $j = 1,2$, and $E_X\{j\}$, $E_\tau\{j\}$; $j = 1,2$, together with the values $\delta_g\{j\}$; $j = 1,2$, and $\delta_X\{j\}$, $\delta_\tau\{j\}$; $j = 1,2$.

l	h_1	h_2	h	E_h	δ_h
0.0	0.864 086 763 083	0.859 360 128 944	1.723 659 400 858	2.1 E-04	-
0.5	1.364 089 626 994	1.359 560 053 306	2.723 658 669 419	9.0 E-06	2.013
1.0	1.864 089 632 342	1.859 568 647 619	3.723 658 668 053	3.9 E-07	2.001
1.5	2.364 089 632 352	2.359 569 018 929	4.723 658 668 050	1.7 E-08	2.000
2.0	2.864 089 632 352	2.859 569 034 974	5.723 658 668 050	7.2 E-10	2.000
2.5	3.364 089 632 352	3.359 569 035 668	6.723 658 668 050	3.1 E-11	2.000

Table 3.2(a)

l	$E_g\{1\}$	$\delta_g\{1\}$	$E_g\{2\}$	$\delta_g\{2\}$
0.0	1.2 E-02	-	1.2 E-02	-
0.5	2.4 E-03	1.022	2.4 E-03	1.018
1.0	5.0 E-04	1.004	5.0 E-04	1.003
1.5	1.0 E-04	1.000	1.0 E-04	1.000
2.0	2.1 E-05	1.000	2.1 E-05	1.000
2.5	4.5 E-06	1.000	4.5 E-06	1.000

Table 3.2(b)

l	$E_X\{1\}$	$\delta_X\{1\}$	$E_X\{2\}$	$\delta_X\{2\}$	$E_\tau\{1\}$	$\delta_\tau\{1\}$	$E_\tau\{2\}$	$\delta_\tau\{2\}$
0.0	4.9 E-04	-	1.1 E-03	-	5.2 E-04	-	4.1 E-04	-
0.5	2.1 E-05	1.999	4.6 E-05	2.001	2.2 E-05	2.005	1.7 E-05	2.002
1.0	9.1 E-07	2.000	2.0 E-06	2.000	9.6 E-07	2.000	7.6 E-07	2.000
1.5	3.9 E-08	2.000	8.6 E-08	2.000	4.2 E-08	2.000	3.3 E-08	2.000
2.0	1.7 E-09	2.000	3.7 E-09	2.000	1.8 E-09	2.000	1.4 E-09	2.000
2.5	7.3 E-11	2.000	1.6 E-10	2.000	7.8 E-11	2.000	6.1 E-11	2.000

Table 3.2(c)

Example 3.3 Q and Q_j; $j = 1, 2$, are defined by (1.3) and (1.4) with

$$\tau_1(x) = 0.75 + 0.2\,\mathrm{sech}^2(2.5x) + l \quad \text{and} \quad \tau_2(x) = x(1-x) + 1 + l.$$

In this case, the condition (1.9) is not fulfilled because $d_2 = 1$.

The numerical results corresponding to the values $l = 0.00(0.25)1.25$ are listed in Tables 3.3(a)-3.3(c). (In this example, the values of h and h_j; $j = 1, 2$, listed in Table 3.3(a) are expected to be correct to eight significant figures.)

l	h_1	h_2	h	E_h	δ_h
0.00	0.815 399 73	1.121 813 26	1.937 329 02	1.2 E-04	-
0.25	1.065 491 74	1.371 813 33	2.437 329 08	2.4 E-05	2.005
0.50	1.315 510 77	1.621 813 33	2.937 329 08	5.0 E-06	2.001
0.75	1.565 514 72	1.871 813 33	3.437 329 08	1.0 E-06	2.000
1.00	1.815 515 54	2.121 813 33	3.937 329 08	2.1 E-07	2.000
1.25	2.065 515 71	2.371 813 33	4.437 329 08	4.5 E-08	2.000

Table 3.3(a)

l	$E_g\{1\}$	$\delta_g\{1\}$	$E_g\{2\}$	$\delta_g\{2\}$
0.00	8.9 E-03	-	8.8 E-03	-
0.25	4.0 E-03	1.027	3.9 E-03	1.017
0.50	1.8 E-03	1.011	1.8 E-03	1.007
0.75	8.1 E-04	1.005	8.1 E-04	1.003
1.00	3.7 E-04	1.002	3.7 E-04	1.001
1.25	1.7 E-04	1.000	1.7 E-04	1.000

Table 3.3(b)

l	$E_X\{1\}$	$\delta_X\{1\}$	$E_X\{2\}$	$\delta_X\{2\}$	$E_\tau\{1\}$	$\delta_\tau\{1\}$	$E_\tau\{2\}$	$\delta_\tau\{2\}$
0.00	8.2 E-04	-	6.8 E-04	-	2.4 E-04	-	2.2 E-04	-
0.25	1.7 E-04	2.003	1.4 E-04	1.999	5.0 E-05	2.007	4.5 E-05	2.000
0.50	3.5 E-05	2.000	2.9 E-05	2.000	1.0 E-05	2.001	9.4 E-06	2.000
0.75	7.3 E-06	2.000	6.1 E-06	2.000	2.2 E-06	2.000	1.9 E-06	2.000
1.00	1.5 E-06	2.000	1.3 E-06	2.000	4.5 E-07	2.000	4.0 E-07	2.000
1.25	3.2 E-07	2.000	2.6 E-07	2.000	9.4 E-08	2.000	8.4 E-08	2.000

Table 3.3(c)

The numerical results of Tables 3.1-3.3 indicate clearly that in the three examples considered above

$$\delta_h = 2, \quad \delta_g\{j\} = 1;\ j = 1,2, \quad \text{and} \quad \delta_X\{j\},\ \delta_\tau\{j\} = 2; \quad j = 1,2.$$

Thus, the numerical results provide experimental support for the Conjectures 3.1 and 3.2, which were made in Remarks 5.2 and 5.4 of [14].

Acknowledgement. One of us (NSS) wishes to thank the State Scholarships Foundation of Greece for their financial support.

References

[1] F. Bowman, *Introduction to Elliptic Functions*, English University Press, London, 1953.

[2] N.V. Challis and D.M. Burley, *A numerical method for conformal mapping*, IMA J. Numer. Anal. **2** (1982), 169-181.

[3] D. Gaier, *Konstruktive Methoden der konformen Abbildung*, Springer, Berlin, 1964.

[4] D. Gaier, *Ermittlung des konformen Moduls von Vierecken mit Differenzenmethoden*, Numer. Math. **19** (1972) 179-194.

[5] D. Gaier, *Determination of conformal modules of ring domains and quadrilaterals*, Lecture Notes in Mathematics 399, Springer, New York, 1974, 180-188.

[6] D. Gaier, *Capacitance and the conformal module of quadrilaterals*, J. Math. Anal. Appl. **70** (1979), 236-239.

[7] D. Gaier, *On an area problem in conformal mapping*, Results in Mathematics, **10** (1986), 66-81

[8] D. Gaier and N. Papamichael, *On the comparison of two numerical methods for conformal mapping*, IMA J. Numer. Anal. **7** (1987), 261-282.

[9] P.Henrici, *Applied and Computational Complex Analysis*, Vol. III, Wiley, New York, 1986.

[10] L.H. Howell and L.N. Trefethen, *A modified Schwarz-Christoffel transformation for elongated regions*, Numerical Analysis Report 88-5, Dept. of Maths, Massachusetts Institute of Technology, Cambridge, Mass., 1988.

[11] C.D. Mobley and R.J. Stewart, *On the numerical generation of boundary-fitted orthogonal curvilinear coordinate systems*, J. Comput. Phys. 34 (1980), 124-135.

[12] N. Papamichael, *Numerical conformal mapping onto a rectangle with applications to the solution of Laplacian problems*, J. Comput. Appl. Math. **28** (1989) 63-83.

[13] N. Papamichael, C.A. Kokkinos and M.K. Warby, *Numerical techniques for conformal mapping onto a rectangle*, J. Comput. Appl. Math. **20** (1987), 349-358.

[14] N. Papamichael and N.S. Stylianopoulos, *A domain decomposition method for conformal mapping onto a rectangle*, Tech. Report TR/04/89, Dept. of Math. and Stat., Brunel University, 1989, (to appear in: Constr. Approx.).

[15] A. Seidl and H. Klose, *Numerical conformal mapping of a towel-shaped region onto a rectangle*, SIAM J. Sci. Stat. Comput. **6** (1985), 833-842.

[16] L.N. Trefethen, Ed., *Numerical Conformal Mapping*, North-Holland, Amsterdam, 1986; reprinted from: J. Comput. Appl. Math. **14** (1986).

[17] J.J. Wanstrath, R.E. Whitaker, R.O. Reid, A.C. Vastano, *Storm surge simulation in transformed co-ordinates*, Tech. Report 76-3, U.S. Coastal Engineering Research Center, Fort Belvoir, Va. 1976.

Received: August 1, 1989.

Computational Methods and Function Theory
Proceedings, Valparaíso 1989
St. Ruscheweyh, E.B. Saff, L. C. Salinas, R.S. Varga (*eds.*)
Lecture Notes in Mathematics **1435**, pp. 171–176

Planar Harmonic Mappings[1]

Glenn Schober
Department of Mathematics, Indiana University
Bloomington, IN 47405, USA

1. Introduction

Harmonic mappings occur in various frameworks and generalities. For this article they are simply complex-valued harmonic functions $f = u + iv$ of the complex variable z, that are one-to-one and orientation-preserving. It is our purpose to show where these mappings occur, to give a brief survey of some areas of study, and to mention several open problems. We begin with some examples, discuss a connection with minimal surfaces, touch on univalent function theory, and conclude with the mapping theory.

2. An example

Let $f(z) = z - 1/\bar{z}$ in $\Delta = \{z : |z| > 1\}$. Then f is clearly harmonic, and from the polar representation $f(re^{i\theta}) = (r - 1/r)e^{i\theta}$ we see that f maps concentric circles in Δ onto concentric circles in $\mathbb{C}\backslash\{0\}$. It is not possible to map Δ conformally, or even quasiconfomally, onto $\mathbb{C}\backslash\{0\}$. So a harmonic mapping may not preserve conformal type.

More generally, it is not much of an exercise to show that the functions

$$f(z) = z - 1/\bar{z} + A\log|z|$$

are also harmonic mappings of Δ onto $\mathbb{C}\backslash\{0\}$ when $|A| \leq 2$. We mention them because they are extremal functions for several problems [7, sect. 3]. For example, the diameter of $\mathbb{C}\backslash f(\Delta)$ is a minimum!

3. More examples and a conjecture

Consider mappings of the annulus $A_\rho = \{z : \rho < |z| < 1\}$ given by $f_t(z) = tz + (1-t)/\bar{z}$. In polar coordinates this is $f_t(re^{i\theta}) = (tr + \frac{1-t}{r})e^{i\theta}$. Again concentric circles are mapped onto concentric circles. We restrict $1/(1+\rho^2) \leq t \leq 1/(1-\rho^2)$ so that

[1]This work was supported in part by a grant from the National Science Foundation (USA).

$tr + \dfrac{1-t}{r}$ is an increasing function of r and nonnegative. Then f_t maps A_ρ onto an annulus $A_\sigma = \{z : \sigma < |z| < 1\}$ and $0 \leq \sigma \leq 2\rho/(1+\rho^2)$.

Now suppose that $f = u + iv$ is an arbitrary harmonic mapping of the annulus A_ρ onto an annulus A_σ. Then $U = u + 1$ is a positive harmonic function to which we shall apply Harnack's inequality in the following form:

Harnack's inequality. *If K is a compact subset of a domain D, then there exists a constant $M = M(K, D)$ such that $U(z) \leq MU(\zeta)$ for all $z, \zeta \in K$ and all positive harmonic functions U in D.*

To be specific, choose for K any circle in A_ρ about the origin. Then for topological reasons $f(K)$ surrounds the origin in A_σ and there are points $z, \zeta \in K$ such that $u(z)$ is to the right of σ and $u(\zeta)$ is to the left of $-\sigma$. That is, we have

$$\sigma + 1 < u(z) + 1 \leq M[u(\zeta) + 1] < M[-\sigma + 1],$$

and this implies $\sigma < (M-1)/(M+1)$. In other words, σ is bounded away from 1 no matter what f is. On the other hand, the examples at the beginning of this section show that σ can be as small as zero. This is a surprising dichotomy. The image annulus cannot be too thin, but it can degenerate to a punctured disk.

These examples and the cute little proof were given by J.C.C. Nitsche [9] in 1962. He conjectures that $\sigma \leq 2\rho/(1+\rho^2)$ for all harmonic mappings of A_ρ onto A_σ. It is still an intriguing open problem. Try it!

4. A connection with minimal surfaces

Let $\mathcal{S}$ denote a nonparametric minimal surface in $\mathbb{R}^3$ that lies above a simply-connected domain Ω in the complex plane, $\Omega \neq \mathbb{C}$. That is, $\mathcal{S} = \{(u, v, F(u, v)) : u + iv \in \Omega\}$ where F satisfies the minimal surface equation

$$\left[1 + \left[\frac{\partial F}{\partial v}\right]^2\right] \frac{\partial^2 F}{\partial u^2} - 2\left[\frac{\partial F}{\partial u}\right]\left[\frac{\partial F}{\partial v}\right] \frac{\partial^2 F}{\partial u \partial v} + \left[1 + \left[\frac{\partial F}{\partial u}\right]^2\right] \frac{\partial^2 F}{\partial v^2} = 0.$$

This equation simply states that the mean curvature (the sum of the two principal curvatures) is zero at each point of $\mathcal{S}$. Although $\mathcal{S}$ is a nonparametric surface, that is, the graph of the function F, $\mathcal{S}$ admits a reparametrization as a parametric surface $\mathcal{S} = \{(u(z), v(z), F(u(z), v(z)) : z \in \mathbb{D}\}$ so that each coordinate function is harmonic. This defines a one-to-one mapping from $\mathbb{D}$ onto $\mathcal{S}$, and its projection down to Ω is a harmonic mapping $f(z) = u(z) + iv(z)$ from $\mathbb{D}$ onto Ω. It is no loss of generality to assume that f is orientation-preserving.

Here is the strategy. Use knowledge about harmonic mappings of $\mathbb{D}$ onto Ω to gain information about nonparametric minimal surfaces $\mathcal{S}$ that lie over Ω.

In 1952, E. Heinz [4] showed that the Fourier coefficients of a harmonic mapping f of $\mathbb{D}$ onto the disk $\Omega_R = \{w : |w| < R\}$ with $f(0) = 0$ satisfy $2(|c_{-1}|^2 + |c_1|^2) > \dfrac{3R^2}{4\pi}$.

As a consequence, this led him to the estimate $|K| \leq \frac{4\pi^3}{3R^2}$ for the gaussian curvature K (product of principal curvatures) at the point above the origin of any nonparametric minimal surface $\mathcal{S}$ that lies over Ω_R. Although the curvature estimate is not sharp, it has a very important consequence. Suppose that $\mathcal{S}$ lies above the entire plane $\mathbb{C}$. Then we may let $R \to \infty$ in the curvature estimate to conclude that $K = 0$. After a translation we find that the sum and product of the principal curvatures are zero at every point. This implies that $\mathcal{S}$ is a plane. In this fashion Heinz gave a new proof of the following:

Bernstein's theorem. *The only nonparametric minimal surfaces that lie over (are parametrized by) the entire plane are themselves planes.*

More recently, R.R. Hall [3] obtained the sharp estimate $2(|c_{-1}|^2 + |c_1|^2) \geq \frac{27R^2}{2\pi^2}$ for what became known as the Heinz constant. Based on it the estimate above for the gaussian curvature can be improved to $|K| \leq \frac{16\pi^2}{27R^2}$, which unfortunately is still not sharp. We shall explain why later, but the basic reason is that there are more harmonic mappings under consideration than nonparametric minimal surfaces. That is, there is no such surface $\mathcal{S}$ that corresponds to the extremal harmonic mapping for the Heinz constant. The sharp estimate for the gaussian curvature is still an interesting open problem.

Various estimates for the gaussian curvature can be found in the book by R. Osserman [10]. Sharp estimates in case Ω is a half-plane, strip, or slit-plane can be found in [8].

The statement that f is harmonic on $\mathbb{D}$ can be written as $f_{z\bar{z}} = 0$ in $\mathbb{D}$. This implies that f_z is analytic and $f_{\bar{z}}$ is anti-analytic. In other words, we have $f = h + \bar{g}$ where h and g are analytic in $\mathbb{D}$. The function $a = \overline{f_{\bar{z}}}/f_z = g'/h'$ has special significance. If f is orientation-preserving, then its Jacobian determinant $J_f = |f_z|^2 - |f_{\bar{z}}|^2$ is nonnegative. As a consequence, the function a is analytic and $|a(z)| \leq 1$. In fact, we have $|a(z)| < 1$, for otherwise, a would be a constant of modulus one and J_f would vanish identically. On each compact subset a is bounded away from one, and so f is locally quasiconformal. However, a harmonic mapping need not be quasiconformal since its distortion may be unbounded at the boundary.

In order for f to derive from a nonparametric minimal surface, the function a must be the square of an analytic function. That is, zeros of a must be of even order. This is not the case for the extremal function for the Heinz constant. When it is defined, the function $\sqrt{a}$ is called the Weierstrass-Enneper function, and $i/\sqrt{a}$ turns out to be the stereographic projection of the Gauss map of the surface $\mathcal{S}$.

5. Univalent harmonic functions

This is an area pioneered by J.G. Clunie and T. Sheil-Small [2]. Let S_H^0 denote the class of one-to-one harmonic orientation-preserving mappings $f = h + \bar{g}$ where h and g are analytic in $\mathbb{D}$ and $h(z) = z + \sum_{n=2}^{\infty} a_n z^n$, $g(z) = \sum_{n=2}^{\infty} b_n z^n$. We shall mention only a few results and compare them with the familiar schlicht class $S = \{f \in S_H^0 : g \equiv 0\}$.

If $f \in S$, then the Koebe one-quarter theorem asserts that the disk $\{w : |w| < 1/4\}$ is contained in the image $f(\mathbb{D})$. For $f \in S_H^0$, Clunie and Sheil-Small have the comparable result that the disk $\{w : |w| < 1/16\}$ is contained in $f(\mathbb{D})$. They conjecture that $1/16$ should be replaced by $1/6$.

If $f \in S$, then the de Branges theorem asserts that the coefficients satisfy $|a_n| \leq n$. For $f \in S_H^0$ the coefficient conjectures are $|a_n| \leq \frac{1}{6}(n+1)(2n+1)$ and $|b_n| \leq \frac{1}{6}(n-1)(2n-1)$, $n \geq 2$. There is a function in S_H^0 that has precisely these numbers as coefficients. It maps $\mathbb{D}$ onto the whole plane with a radial slit from $-1/6$ to infinity. Here, too, most of these inequalities are still conjectures.

For $f \in S_H^0$, the corresponding function $a = g'/h'$ is bounded by one and vanishes at the origin. Therefore Schwarz's lemma implies $|2b_2| = |a'(0)| \leq 1$. In this way the coefficient conjecture $|b_2| \leq 1/2$ is verified. What about the conjecture $|a_2| \leq 5/2$? In [2] they prove that $|a_2| < 12{,}173$ (!). More recently, Sheil-Small has reduced this estimate to $|a_2| < 57$. There is a lot of room for improvement, and very little is known about the higher coefficients.

Clunie and Sheil-Small [2] have many sharp results in case $f \in S_H^0$ is convex, convex in one direction, or close-to-convex.

6. A mapping theorem

Here is the problem:

Mapping problem. Let Ω be a simply-connected domain in $\mathbb{C}$, $\Omega \neq \mathbb{C}$, and let a be analytic in $\mathbb{D}$ and satisly $|a(z)| < 1$ for $z \in \mathbb{D}$. Does there exist a one-to-one harmonic mapping f of $\mathbb{D}$ onto Ω that satisfies the differential equation $\overline{f_{\bar{z}}} = af_z$?

The case $a \equiv 0$ is taken care of by the Riemann mapping theorem, and so we assume a is not identically zero.

First we note that a smooth solution of the differential equation $\overline{f_{\bar{z}}} = af_z$ will necessarily be harmonic. To see this, take the $\bar{z}$-derivatives of both sides to obtain $|\overline{f_{\bar{z}z}}| = |a||f_{z\bar{z}}|$. Since $|a| < 1$, we must have $f_{z\bar{z}} = 0$.

The following example shows that some caution is in order.

Example. Let $\Omega = \mathbb{D}$ and $a(z) = z$. If we write $f = h + \bar{g}$ where $h(z) = \sum_{n=0}^{\infty} a_n z^n$ and $g(z) = \sum_{n=1}^{\infty} b_n z^n$, then the differential equation becomes $g' = zh'$. It implies that $(n+1)b_{n+1} = na_n$ for $n \geq 0$. Therefore the area of $f(\mathbb{D})$ is

$$A = \iint_{\mathbb{D}} (|f_z|^2 - |f_{\bar{z}}|^2)dxdy = \pi \sum_{n=1}^{\infty} n[|a_n|^2 - |b_n|^2] = \pi \sum_{n=0}^{\infty} \left[\frac{n}{n+1}\right] |a_n|^2$$

and the mean

$$M_2 = \frac{1}{2\pi} \int_0^{2\pi} |f(e^{i\theta})|^2 d\theta = \sum_{n=0}^{\infty} [|a_n|^2 + |b_{n+1}|^2] = \sum_{n=0}^{\infty} \left[1 + \left[\frac{n}{n+1}\right]^2\right] |a_n|^2.$$

Since $2\left[\frac{n}{n+1}\right] < 1 + \left[\frac{n}{n+1}\right]^2$, we find that $A < \frac{\pi}{2} M_2$. If $f(\mathbb{D}) \subset \mathbb{D}$, then $M_2 \leq 1$ and the area of $f(\mathbb{D})$ is less than $\pi/2$. Since the area of $\mathbb{D}$ is π, the function f cannot map $\mathbb{D}$ onto $\mathbb{D}$.

This example shows that the answer to the mapping problem as stated can be no. The following is the best positive statement that we know at present. (cf. [6]).

Mapping Theorem [6]. *Let Ω be a bounded simply-connected domain with locally connected boundary. Fix $w_0 \in \Omega$, and let a be analytic in* D *and satisfy $|a(z)| < 1$. Then there exists a univalent, harmonic, orientation-preserving mapping f with the following properties:*

(a) f maps D *into Ω and $f(0) = w_0$, $f_z(0) > 0$.*

(b) f satisfies $\overline{f_{\bar{z}}} = af_z$.

(c) There exists a countable set $E \subset \partial \mathbb{D}$ such that

(i) the unrestricted limit $f(e^{i\theta})$ exists, is continuous, and belongs to $\partial\Omega$ for $e^{i\theta} \in \partial\mathbb{D}\backslash E$;

(ii) the one-sided limits $f(e^{i(\theta+0)})$ and $f(e^{i(\theta-0)})$ exist, are different, and belong to $\partial\Omega$ for $e^{i\theta} \in E$;

(iii) the cluster set of f at $e^{i\theta} \in E$ is the straight-line segment joining $f(e^{i(\theta+0)})$ to $f(e^{i(\theta-0)})$.

The theorem does not and cannot claim that $f(\mathbb{D}) = \Omega$. However, the boundary values of f exist and lie on $\partial\Omega$ with only countably many exceptions. We emphasize that the function $|a|$ is permitted to tend to one at the boundary of D. In that case the mapping f is only locally quasiconformal, the corresponding differential equation is no longer uniformly elliptic, and one may expect pathology at the boundary. In general, the uniqueness of f is still an open question.

What happens in the case of our example $\Omega = \mathbb{D}$, $a(z) = z$? The function

$$f(z) = \frac{1}{2\pi}\int_{-\pi}^{\pi} g(e^{i\theta})\mathrm{Re}\{(e^{i\theta}+z)/(e^{i\theta}-z)\}d\theta$$

with $g(e^{i\theta}) = e^{2\pi ki/3}$ for $(2k-1)\pi/3 < \theta < (2k+1)\pi/3$ and k = -1, 0, 1 satisfies $\overline{f_{\bar{z}}} = zf_z$ and all other conclusions of the mapping theorem. It maps D onto a triangle with vertices at $e^{-2\pi i/3}$, 1, $e^{2\pi i/3}$. Except for three points on $\partial\mathbb{D}$, these are the boundary values of f. The cluster sets of f at those three points account for the edges of the triangle. In this case we know that the mapping is unique [5, sect. 6].

7. Constructive approximation

It is of interest actually to construct the mappings from the theorem in the previous section. This has been carried out by D. Bshouty, N. Hengartner, and W. Hengartner [1] in case Ω is strictly starlike and $\|a\|_\infty < 1$. Under these hypotheses they show also that the mapping is unique.

References

[1] D. Bshouty, N. Hengartner, W. Hengartner, *A constructive method for starlike harmonic mappings*, Numer. Math. **54** (1988), 167-178.

[2] J. Clunie, T. Sheil-Small, *Harmonic univalent functions*, Ann. Acad. Sci. Fenn. Ser. AI **9** (1984), 3-25.

[3] R.R. Hall, *On an inequality of E. Heinz*, J. Analyse Math. **42** (1982/83), 185-198.

[4] E. Heinz, *Über die Lösungen der Minimalflächengleichung*, Nachr. Akad. Wiss. Göttingen, Math.-Phys. Kl., (1952), 51-56.

[5] W. Hengartner, G. Schober, *On the boundary behavior of orientation-preserving harmonic mappings*, Complex Variables Theory Appl. **5** (1985), 181-192.

[6] W. Hengartner, G. Schober, *Harmonic mappings with given dilatation*, J. London Math. Soc. **33** (1986), 473-483.

[7] W. Hengartner, G. Schober, *Univalent harmonic functions*, Trans. Amer. Math. Soc. **299** (1987), 1-31.

[8] W. Hengartner, G. Schober, *Curvature estimates for some minimal surfaces*, in Complex Analysis, Articles Dedicated to Albert Pfluger on the Occasion of His 80th Birthday, J. Hersch and A. Huber Eds., Birkhäuser, 1988, 87-100.

[9] J.C.C. Nitsche, *On the module of doubly-connected regions under harmonic mappings*, Amer. Math. Monthly **69** (1962), 781-782.

[10] R. Osserman, *A Survey of Minimal Surfaces*, Dover, 1986.

Received: June 1, 1989.

Computational Methods and Function Theory
Proceedings, Valparaíso 1989
St. Ruscheweyh, E.B. Saff, L. C. Salinas, R.S. Varga (*eds.*)
Lecture Notes in Mathematics **1435**, pp. 177–190

Extremal Problems for Non-vanishing H^p Functions

T.J. Suffridge

Department of Mathematics, University of Kentucky
Lexington, KY 40506, USA

1. Introduction

The family $\mathcal{P}$ of functions that are analytic in the unit disk $D = \{|z| < 1\}$ and satisfy $P(0) = 1$ and $\mathrm{Re}[P(z)] > 0$ when $z \in D$ are well understood ([2], [6]). The family $\mathcal{B}$ of bounded analytic functions that never assume the value 0 are related to $\mathcal{P}$ as follows. Let $f \in \mathcal{B}$. Then $f(z) = \gamma e^{-tP(z)}$ where $|\gamma| = 1$, $t \geq 0$ and $P \in \mathcal{P}$. With this simple relationship between the classes $\mathcal{P}$ and $\mathcal{B}$, it is surprising that the problem: Find $A_n = \max_{f \in \mathcal{B}} |a_n|$ for fixed $n = 1, 2, \ldots$ (where $f(z) = a_0 + a_1 z + \cdots + a_n z^n + \cdots$) is so difficult even for relatively small n. The problem is trivial for $n = 1$, it can be solved rather easily for $n = 2$ but it is much more difficult for $n = 3$, [3] and $n = 4$ [5]. The problem has not been solved for $n \geq 5$. The solution conjectured by Krzyz [4] is that $A_n = 2/e$ and that $|a_n| < 2/e$ unless

$$f(z) = ce^{(\gamma z^n - 1)/(\gamma z^n + 1)}, \quad |c| = |\gamma| = 1. \tag{1}$$

This conjecture is indeed true for $n = 1, 2, 3, 4$.

A rather natural extension of the above question is to assume

$$f \in H^p, \quad M_p(f) = \sup_{r<1} \left[\frac{1}{2\pi} \int_0^{2\pi} |f(re^{i\theta})|^p d\theta \right]^{1/p} \leq 1$$

and f is non-vanishing and to again ask for $A_n(p) = \sup |a_n|$ the sup being taken over the family described above. Hummel, et. al. [3] conjectured that $A_n(p) = (2/e)^{1/q}$, $\frac{1}{p} + \frac{1}{q} = 1$, $1 < p$ and that the only extremal functions are

$$f(z) = c\frac{(1 + \gamma z^n)^{2/p}}{2^{1/p}} e^{(1-\frac{1}{p})(\gamma z^n - 1)/(\gamma z^n + 1)} \tag{2}$$

where $|c| = |\gamma| = 1$.

Brown [1] has verified this conjecture for $n = 1$ and for other values of n with the added hypothesis that certain coefficients are 0. Hummel, et. al. [3] observed that for $p = 1$, it is easy to see that $A_n(1) = 1$. This follows from the fact that

$$a_n = \frac{1}{2\pi}\int_0^{2\pi} \frac{1}{r^n e^{in\theta}} f(re^{i\theta})d\theta,$$

so that

$$|a_n| \leq \frac{1}{2\pi}\int_0^{2\pi} \frac{1}{r^n}|f(re^{i\theta})|d\theta \leq \frac{1}{r^n}$$

for all $r < 1$, assuming $f \in H^1$, $||f||_1 \leq 1$. However, $|a_n| = 1$ for every polynomial

$$\sum_{j=0}^{2n} a_j z^j = c(\prod_{j=1}^{n}(1 + e^{i\alpha_j}z))^2 = c(\sum_{j=0}^{n} b_j z^j)^2,$$

where $c = 1/\sum_{j=1}^{n} |b_j|^2$.

In this paper we study continuous linear functionals on the family N_p of functions in H^p, $0 < p$, with the properties $||f||_p \leq 1$, and $f(z) \neq 0$ when $z \in D$ or $f \equiv 0$. For each p, N_p is compact in the topology of uniform convergence on compact sets. We let $\mathcal{L}$ denote the family of complex valued continuous linear functionals on the family of functions that are analytic in D with the topology of uniform convergence on compact sets. For a given $L \in \mathcal{L}$, we study $\max_{f \in N_p} \text{Re}L(f)$ which is the same as $\max_{f \in N_p} |L(f)|$ because $F \in N_p$ if and only if $\gamma f \in N_p$ for all γ, $|\gamma| = 1$. Our main theorem is the following.

Theorem 1. *Suppose $L \in \mathcal{L}$ is non-trivial and $f \in N_p$ satisfies $\text{Re}L(f) = \max_{g \in N_p}[\text{Re}L(g)]$, $0 < p < \infty$. Then there is a function $h \in H^2$ that extends to the closed disk $\bar{D}$ so that —h— is continuous, $||h||_2 = 1$ and having the following properties.*

(i) $\text{Re}\left[L\left(\frac{1 + ze^{i\alpha}}{1 - ze^{i\alpha}}f\right)\right] = L(f)|h(e^{-i\alpha})|^2.$

(ii) $f(z) = (h(z))^{2/p}I(z)$, *where I is an inner function with $I = e^{-P}$,*

$$P(z) = \int_0^{2\pi} \frac{1 + ze^{i\alpha}}{1 - ze^{i\alpha}} d\mu(\alpha),$$

where μ is a positive measure (or 0) with support contained in the zero set of $h(e^{i\theta})$.

(iii) If L depends only on the coefficients $a_0, a_1, ..., a_n$ of an analytic function $\sum a_j z^j$, then h is a polynomial of degree $\leq n$ and μ consists of at most n point masses.

In case $p = \infty$ (i), (ii) and (iii) hold for at least one extremal function f.

In order to see how one may apply theorem 1, we prove theorem 2 below.

Theorem 2. *Suppose $f \in N_p$, $a_0 > 0$ and $a_1 \geq 0$ (this can be accomplished by replacing f by $\gamma f(ze^{i\alpha})$ for appropriate α and γ), then* *(i)* $a_1 \leq (2/e)^{1-1/p}$ *when* $p \geq 1$ *with equality if and only if*

$$f(z) = 2^{-1/p}(1+z)^{2/p}e^{-(1-1/p)(1-z)/(1+z)}$$

and

(ii) $a_1 \leq p^{-1/2}(2-p)^{1/p-1/2}2^{1-1/p}$ *when* $0 < p < 1$ *with equality if and only if*

$$f(z) = (1+\rho^2)^{-1/p}(1+\rho z)^{2/p}$$

where $\rho^2 = p/(2-p)$.

Proof. Choose $L(f) = a_1$. By theorem 1, (iii), h is a polynomial of degree zero or 1. If degree $h = 0$, using (ii) $f \equiv 1$ so $L(f) = 0$ which is clearly not maximal. Therefore degree $h = 1$. We may assume f has real coefficients because $f \in H_p$ implies $g(z) = (f(z)\overline{f(\bar{z})})^{1/2} \in H_p$ while g has real coefficients and $g(z) = a_0 + a_1 z + \cdots$, $f(z) = a_0 + a_1 z + a_2 z^2 + \cdots$, i.e. the first two coefficients agree. Using (i) we see that $a_1 + a_0 e^{i\alpha} + a_0 e^{-i\alpha} = a_1(c_0 + c_1 e^{-i\alpha})(\bar{c}_1 e^{i\alpha} + \bar{c}_0)$. Thus c_0 and c_1 can be assumed to be positive and

$$a_0 + a_1 z + a_0 z^2 = a_1(c_0 z + c_1)(c_1 z + c_0). \tag{3}$$

Now suppose μ is not 0. Then h has a zero on $|z| = 1$ and hence $c_1 = c_0 = 1/\sqrt{2}$ so

$$f(z) = 2^{-1/p}(1+z)^{2/p}e^{-t(1-z)/(1+z)} \tag{4}$$

is a possible extremal. Since $a_0 = a_1/2$ by (3), while $a_1/a_0 = 2/p + 2t$ by (4), we conclude $t = 1 - 1/p \geq 0$ and this is a possible extremal only if $p \geq 1$.

If μ is zero then $I(z) \equiv 1$, $c_1 = \rho c_0 \leq c_0$ and $c_0 = 1/\sqrt{1+\rho^2}$. Thus

$$f(z) = (1+\rho^2)^{-1/p}(1+\rho z)^{2/p}, \tag{5}$$

so that $\frac{a_1}{a_0} = \frac{2\rho}{p}$ by (5) and $a_0 = a_1 \frac{\rho}{1+\rho^2}$ by (3). Thus $1 + \rho^2 = \frac{2}{p}\rho^2$ and $\left(\frac{2}{p} - 1\right)\rho^2 = 1$. Since $\rho \leq 1$, this is only possible when $p \leq 1$. The proof is now complete since in this case, $\rho^2 = \frac{p}{2-p}$. ■

Note that the result of theorem 2 was given by Brown [1] by a different method for $p > 1$ while the result for $0 < p < 1$ is new. Further, it is easy to see that Brown's result that if $f \in N_p$, $f(z) = a_0 + a_m z^m + \cdots + a_n z^n + \cdots$ where $m > \frac{n}{2}$, $p > 1$, then $|a_n| \leq (2/e)^{1-\frac{1}{p}}$ is true, as follows. Under the above assumption, the function

$$g(z) = \left(\prod_{j=1}^{n} f_j(z)\right)^{1/n}, \quad f_j(z) = f(ze^{2\pi i(j-1)/n}),$$

is n-fold symmetric (i.e. $g(z) = k(z^n)$ for some $k \in N_p$) and satisfies

$$g(z) = a_0 + a_n z^n + \cdots.$$

It therefore follows that a_n for this restricted class has the same bound as a_1 for the entire family. While it is conjectured that the extreme value for $|a_n|$ in the entire family is identical to that for $|a_1|$ when $p \geq 1$. This is not true for $p < 1$: As we shall see below, $|a_2|$ has the sharp bound $2^{2-2/p}(2+p)^{2/p-1}/p$ (the square of the sharp bound for $|a_1|$) when $p < 1$.

2. Proof of main theorem

Given a non-trivial function $f \in N_p$, we may write

$$f(z) = (h(z))^{2/p} I(z), \tag{6}$$

where $h \in H^2$, $h(z) \neq 0$ when $z \in D$ and $I(z) = e^{-P(z)}$ is an inner function, $h(0) > 0$, $P(0) \geq 0$.

Lemma 1. *Suppose h and k are outer functions in the space H^2 and that $h(0) > 0, k(0) > 0$. If h^* and k^* are the corresponding boundary functions and $|h^*(e^{i\theta})| = |k^*(e^{i\theta})|$ a.e., then $h = k$.*

Proof. By hypothesis,

$$\begin{aligned} h(z) &= \exp\left\{\frac{1}{2\pi}\int_{-\pi}^{\pi} \frac{e^{it}+z}{e^{it}-z} \log|h^*(e^{it})| dt\right\} \\ &= \exp\left\{\frac{1}{2\pi}\int_{-\pi}^{\pi} \frac{e^{it}+z}{e^{it}-z} \log|k^*(e^{it})| dt\right\} \\ &= k(z). \end{aligned}$$

■

Lemma 2. *If $h(z) = \sum_{k=0}^{\infty} c_k z^k \in H^2$, $h(z) \neq 0$ in D then for $0 \leq \epsilon \leq 1$, the function*

$$h_\epsilon(z) = (1 + \epsilon e^{i\beta} z^n) h(z) \left[\sum_{k=0}^{\infty} |c_k|^2 (1+\epsilon^2) + 2\epsilon \mathrm{Re} \sum_{k=0}^{\infty} c_k \bar{c}_{k+n} e^{i\beta}\right]^{-1/2}$$

has the same properties and $||h_\epsilon||_2 = 1$.

Proof. Use the fact that for $\sum_{k=0}^{\infty} a_k z^k = f(z) \in H^2$, $||f||_2^2 = \sum_{k=0}^{\infty} |a_k|^2$ and

$$(1 + \epsilon e^{i\beta} z^n) h(z) = \sum_{k=0}^{n-1} c_k z^k + \sum_{k=0}^{\infty} (c_{k+n} + \epsilon e^{i\beta} c_k) z^{k+n}.$$ ■

Lemma 3. *If h and h_ϵ are related as in Lemma 2, then*

$$\left.\frac{d}{d\epsilon} h_\epsilon(z)\right|_{\epsilon=0} = \left(e^{i\beta} z^n - \mathrm{Re}\left(e^{i\beta} \frac{\sum_{k=0}^{\infty} c_k \bar{c}_{k+n}}{||h||_2^2}\right)\right) \frac{h}{||h||_2}.$$

Proof. A straightforward differentiation. ■

Lemma 4. *If $L \in \mathcal{L}$ is a non-trivial linear functional and $f \in N_p$ has the property $\mathrm{Re}L(f) \geq \mathrm{Re}L(g)$ for all $g \in N_p$, $0 < p < \infty$, where $f = h^{2/p} I$, $h \in H^2$ and I is inner, then $||h||_2 = 1$ and*

$$L(z^n f) = L(f) \sum_{k=0}^{\infty} c_k \bar{c}_{k+n} \text{ for } n = 0, 1, \cdots. \tag{7}$$

Proof. Clearly $||f||_p = ||h||_2$ and $||f||_p < 1$ implies $\rho f \in N_p$ for some $\rho > 1$ so that $\text{Re}L(\rho f) = \rho\text{Re}L(f) > \text{Re}L(f)$ (because L is non-trivial and the assumed extremal f satisfies $\text{Re}L(f) > 0$). Thus, (7) holds for $n = 0$. Now consider $f_\epsilon = (h_\epsilon)^{2/p}I$, h_ϵ as in Lemma 2. Then $\text{Re}L(f_\epsilon) \le \text{Re}L(f)$ and hence

$$\text{Re}L\left(\left.\frac{df_\epsilon}{d\epsilon}\right|_{\epsilon=0}\right) \le 0.$$

We have

$$\left.\frac{df_\epsilon}{d\epsilon}\right|_{\epsilon=0} = \frac{2}{p}(h_\epsilon)^{2/p-1}I\left.\frac{dh_\epsilon}{d\epsilon}\right|_{\epsilon=0} = \frac{2}{p}f\left(e^{i\beta}z^n - \text{Re}e^{i\beta}\sum_{k=0}^{\infty}c_k\bar{c}_{k+n}\right).$$

Therefore

$$\text{Re}\left\{e^{i\beta}L(z^n f) - \text{Re}\ e^{i\beta}\sum_{k=0}^{\infty}c_k\bar{c}_{k+n}L(f)\right\} \le 0.$$

That is, $\text{Re}\{e^{i\beta}(L(z^n f) - \sum_{k=0}^{\infty}c_k\bar{c}_{k+n}L(f))\} \le 0$. Since β is arbitrary, the lemma now follows. ∎

Note that the expression

$$\sum_{k=0}^{\infty}c_k\bar{c}_{k+n}$$

is

$$\frac{1}{2\pi}\int_0^{2\pi}e^{in\theta}h^*(e^{i\theta})\overline{h^*(e^{i\theta})}d\theta.$$

This is the coefficient of $e^{-in\theta}$ in the Fourier series for $|h^*(e^{i\theta})|^2$ and we write it as $\langle|h|^2, \bar{z}^n\rangle = \langle z^n, |h|^2\rangle$. For $0 < \rho < 1$, $0 < p < \infty$ and f extremal for L (i.e. $\text{Re}L(f) \ge \max_{g\in N_p}\text{Re}L(g)$), we have

$$L\left(\frac{1+z(\rho e^{i\alpha})}{1-z(\rho e^{i\alpha})}f\right) = L(f) + 2\sum_{n=1}^{\infty}L(z^n f)\rho^n e^{in\alpha} = L(f) + 2\sum_{n=1}^{\infty}L(f)\langle z^n, |h|^2\rangle\rho^n e^{in\alpha}.$$

Using the continuity of L and taking real parts, we obtain

$$\begin{aligned}\text{Re}L\left(\frac{1+ze^{i\alpha}}{1-ze^{i\alpha}}f\right) &= \text{Re}(L(f) + 2\sum_{n=1}^{\infty}L(f)\langle z^n, |h|^2\rangle e^{in\alpha}) \\ &= L(f)\left(1 + \sum_{n=1}^{\infty}\langle z^n, |h|^2\rangle e^{in\alpha} + \langle\bar{z}^n, |h|^2\rangle e^{-in\alpha})\right) \\ &= L(f)|h^*(e^{-i\alpha})|^2\end{aligned}$$

is continuous. This is (i) of the main theorem for $0 < p < \infty$.

To obtain (ii), note that everything is clear except that the support of μ is contained in the zero set of $h(e^{i\theta})$. To see this, note that $\text{Re}L(fI^\epsilon)$ has a maximum at $\epsilon = 0$. Thus, $\text{Re}L(f\log I) = 0$. That is

$$\operatorname{Re} L\left(f\int_0^{2\pi}\frac{1+ze^{it}}{1-ze^{it}}d\mu\right)=L(f)\int_0^{2\pi}|h(e^{it})|^2d\mu=0,$$

and (ii) follows.

In case L depends only on the coefficients $a_0, a_1, \cdots, a_n$, then $L(z^m f) = 0$ when $m > n$. This implies $\langle z^m, |h|^2\rangle = 0$ for $m > n$ and by a result of Fejer, (since $|h|^2$ is a non-negative real trigonometric polynomial of degree $\leq n$), h is a polynomial of degree n. To see this, note that

$$|h(e^{i\theta})|^2=\sum_{k=-n}^{n}d_k e^{ik\theta},\quad d_{-k}=\bar{d}_k.$$

With

$$K(z)=\sum_{k=-n}^{n}d_k z^{k+n},$$

we have

$$e^{-in\theta}K(e^{i\theta})=|h(e^{i\theta})|^2.$$

Clearly, the zeros of K on $|z| = 1$ are of even multiplicity and since

$$K(z)=z^{2n}\overline{K\left(\frac{1}{\bar{z}}\right)}$$

(i.e., K is self inversive) we have

$$K(z)=c\left(\prod_{j=1}^{s}(1+e^{i\alpha_j}z)^2\right)\prod_{j=s+1}^{n}\left[(1+\rho_j e^{i\alpha_j}z)(1+\frac{1}{\rho_j}e^{i\alpha_j}z)\right],$$

where each $\rho_j < 1$ and $c > 0$.

Set

$$H(z)=K(z)\prod_{j=s+1}^{n}\frac{(1+\rho_j e^{i\alpha_j}z)}{\rho_j\left(1+\frac{1}{\rho_j}e^{i\alpha_j}z\right)}=c'\prod_{j=1}^{n}(1+\rho_j e^{i\alpha_j}z)^2,$$

where $\rho_j = 1,\ \ 1 \leq j \leq s,\ \ \rho_j < 1, j > s,\ \ c' > 0$.

Thus, $Q(z) = \sqrt{c'}\prod_{j=1}^{n}(1 + \rho_j e^{i\alpha_j}z)$ has the property, Q is outer and $|Q(e^{i\theta})|^2 = |h(e^{i\theta})|^2$. This means $Q \equiv h$.

This completes the proof of (iii) in the main theorem.

To prove that (i), (ii) and (iii) hold for some extremal f in case $p = \infty$, let f_p be extremal for L in N_p, $p \geq 2$. Since $N_\infty \subset N_p \subset N_2$, $2 \leq p$, the family $\{f_p\} = \{(h_p)^{2/p}I_p\}$ is locally uniformly bounded. Thus, as $p \to \infty$, there is a sequence $\{p_n\}$ such that $p_n \to \infty$ and $h_{p_n} \to h, I_{p_n} \to I$ uniformly on compact subsets of the disk D. Clearly, $\operatorname{Re}L(f_p) \geq \max_{g\in N_\infty}\operatorname{Re}L(g)$ and hence $\operatorname{Re}L(f) \geq \max_{g\in N_\infty}\operatorname{Re}L(g)$ by continuity of L. The theorem now follows. ∎

3. Applications

Proposition. *If $f \in N_p$ then $|f(r)| \leq \frac{1}{(1-r^2)^{1/p}}$ with equality for*

$$f(z) = \frac{(1-r^2)^{1/p}}{(1-rz)^{2/p}}.$$

Of course one can prove this directly using the fact that for $f \in H^2$, $||f|| \leq 1$,

$$\begin{aligned}
|f(r)| &= \left|\frac{1}{2\pi i}\int_{|z|=\rho}\frac{f(z)}{z-r}dz\right| \\
&= \left|\frac{1}{2\pi}\int_0^{2\pi}\frac{\rho e^{i\theta}f(\rho e^{i\theta})}{\rho e^{i\theta}-r}d\theta\right| \\
&\leq \frac{1}{2\pi}\int_0^{2\pi}\frac{|f(\rho e^{i\theta})|}{|1-\frac{r}{\rho}e^{-i\theta}|}d\theta \\
&\leq \left(\frac{1}{2\pi}\int_0^{2\pi}|f(\rho e^{i\theta})|^2 d\theta\right)^{1/2}\left(\frac{1}{2\pi}\int_0^{2\pi}\frac{1}{|1-\frac{r}{\rho}e^{-i\theta}|^2}d\theta\right)^{1/2} \\
&\leq \frac{1}{\sqrt{1-r^2}},
\end{aligned}$$

while $g \in H^p$, $||g||_p \leq 1$, $g(z) \neq 0$ in D implies $g(z) = (f(z))^{2/p}$ for some $f \in H^2$, $||f|| \leq 1$.

Using Theorem 1, the proof is as follows. For the extremal function f,

$$r^n f(r) = L(z^n f) = \langle z^n, |h|^2\rangle f(r).$$

This implies

$$\begin{aligned}
|h(e^{i\theta})|^2 &= 1 + \textstyle\sum_{n=1}^{\infty} r^n(e^{in\theta}+e^{-in\theta}) \\
&= 1 + \frac{re^{i\theta}}{1-re^{i\theta}} + \frac{re^{-i\theta}}{1-re^{-i\theta}} \\
&= \frac{1-r^2}{|1-re^{i\theta}|^2}.
\end{aligned}$$

We conclude $h(z) = \frac{\sqrt{1-r^2}}{1-rz}$ (up to a constant of modulus 1) and since $h(z) \neq 0$ when $|z| = 1$, $I \equiv 1$. Therefore, $f(z) = \frac{(1-r^2)^{1/p}}{(1-rz)^{2/p}}$ is extremal and $f(r) \leq 1/(1-r^2)^{1/p}$. ∎

Remark 1. Theorem 1 remains true in the restricted class $\{f \in N_p : \quad f(0) > 0\}$ provided the functional L has the property that $L(f) > 0$ for extremal functions f with $\max \mathrm{Re} L(g) = \mathrm{Re} L(f)$ over the restricted family.

Theorem 3. *If $f \in N_p$, and $f(0) > 0$, then*

$$\mathrm{Re}(f(r) - f(0)) \leq 2r/(1-r)\left(\frac{1-r}{2}\right)^{1/p} e^{-t},$$

when $(1+r)(1-r)^{2/p-1} \geq 1$, where

$$t = \frac{1+r}{2r}\log[(1+r)(1-r)^{2/p-1}]$$

and

$$\mathrm{Re}(f(r) - f(0)) \leq \frac{r}{p}\left(\frac{1+2\rho r+\rho^2}{1-r^2}\right)^{1-1/p},$$

when $(1+r)(1-r)^{2/p-1} < 1$, where ρ is a solution of the equation

$$\left(\frac{1+\rho r}{1-r^2}\right)^{2/p-1} = \frac{\rho + r}{\rho}.$$

Remark 2. Note that the result in Theorem 3 is not the correct bound for the entire class N_p. This is easily seen in case $p = \infty$. In this case, since $\mathrm{Re} f(r) \leq 1$, if we require $f(0) > 0$, clearly $\mathrm{Re}(f(r) - f(0)) < 1$. However, if $f(z) = -e^{-tP(z)}$ where $P(0) = 1$ and t is small and positive then $f(0)$ can be near -1 but with r near 1, we may choose P so that $tP(z) = u + v_i, \quad u$ small, $v = \pi$. Then $f(r) - f(0)$ will be nearly 2. The problem $\max_{f \in N_p} \mathrm{Re}(f(r) - f(0))$ is more difficult.

Proof of Theorem 3. If $f \in N_p, \quad f(0) > 0$ then the function $g(z) = (f(z) \cdot \overline{f(\bar{z})})^{1/2} \in N_p, \quad g(0) = f(0) > 0$ and $g(r) = |f(r)| \geq \mathrm{Re} f(r)$. Thus, by Remark 1, Theorem 1 applies to the functional $L(f) = f(r) - f(0)$. Further, the extremal function can be assumed to have real coefficients. We have $r^n f(r) = L(z^n f) = [f(r) - f(0)]\langle z^n, \quad |h|^2\rangle$, so that

$$\begin{aligned}
|h(e^{i\theta})|^2 &= 1 + \sum_{n=1}^{\infty} \frac{f(r)}{f(r)-f(0)} r^n (e^{in\theta} + e^{-in\theta}) \\
&= 1 + \frac{f(r)}{f(r)-f(0)}\left(\frac{re^{i\theta}}{1-re^{i\theta}} + \frac{re^{-i\theta}}{1-re^{-i\theta}}\right) \\
&= \frac{1}{|1-re^{i\theta}|^2}\left(1 + r^2 - r(e^{i\theta} + e^{-i\theta}) + \frac{rf(r)}{f(r)-f(0)}(e^{i\theta} - r + e^{-i\theta} - r)\right) \\
&= \frac{1}{|1-rz|^2}\left|\frac{rf(0)}{f(r)-f(0)} + (1 - r^2 - 2r^2\frac{f(0)}{f(r)-f(0)}z + \frac{rf(0)}{f(r)-f(0)}z^2\right| \\
&= \frac{1}{|1-rz|^2}\frac{K}{|\rho|}|(1+\rho z)(\rho + z)|,
\end{aligned}$$

where ρ is real, $|\rho| \leq 1$, $z = e^{i\theta}$, and $K = \dfrac{rf(0)}{f(r)-f(0)}$. If $|\rho| = 1$, it is easy to check that $\rho \neq -1$ (i.e., the relation

$$1 - r^2 - 2r^2 \frac{f(0)}{f(r)-f(0)} = -\frac{2rf(0)}{f(r)-f(0)}$$

cannot hold), so $|\rho| = 1$ implies $\rho = 1$. In this case, we find that

$$h(z) = \sqrt{K}\frac{1+z}{1-rz} \quad \text{and} \quad f(z) = K^{1/p}\left(\frac{1+z}{1-rz}\right)^{2/p} e^{-t\frac{1-z}{1+z}}.$$

Since $||f||_p = 1$, $K = \dfrac{1-r}{2}$, so that $f(r) - f(0) = \dfrac{2r}{1-r}f(0)$, $f(r) = \dfrac{1+r}{1-r}f(0)$ and hence

$$\left(\frac{1+r}{1-r}\right)^{2/p} e^{-t(1-r)/(1+r)} = \frac{1+r}{1-r}e^{-t}.$$

Thus, $e^{\frac{2tr}{1+r}} = (1+r)(1-r)^{2/p-1}$ and we obtain the value of t given in the theorem. Because $t \geq 0$, this can be extremal only for the values of p for which $(1+r)(1-r)^{2/p-1} \geq 1$ (this clearly includes $p \geq 2$). In case $|\rho| < 1$, we find as before that $\rho > 0$, so that

$$|h(z)| = \frac{1}{|1-rz|^2}\frac{K}{\rho}|(1+\rho z)(\rho + z)| = \frac{1}{|1-rz|^2}\frac{K}{\rho}|1+\rho z|^2,$$

$z = e^{i\theta}$, and we conclude $h(z) = \sqrt{\dfrac{K}{\rho}}\dfrac{1+\rho z}{1-rz}$, $|z| < 1$. Thus, $f(z) = \left(\dfrac{K}{\rho}\right)^{1/p}\left(\dfrac{1+\rho z}{1-rz}\right)^{2/p}$.

Again using $||f||_p = 1$, we get $K = \dfrac{\rho(1-r^2)}{1+2\rho r+\rho^2}$, so

$$f(r) - f(0) = \frac{r(1+2\rho r+\rho^2)}{\rho(1-r^2)}f(0), \quad f(r) = \frac{r+\rho}{\rho}\frac{(1+\rho r)}{1-r^2}f(0).$$

Thus, since $f(0) = \left(\dfrac{K}{\rho}\right)^{1/p}$ we have

$$f(r) - f(0) = \frac{r}{\rho}\left(\frac{1+2\rho r+\rho^2}{1-r^2}\right)^{1-1/p}.$$

Using $f(r) = \dfrac{r+\rho}{\rho}\left(\dfrac{1+\rho r}{1-r^2}\right)f(0)$ we see that ρ must satisfy $\left(\dfrac{1+\rho r}{1-r^2}\right)^{2/p-1} = \dfrac{r+\rho}{\rho}$. This clearly has no solution when $p \geq 2$ and since ρ varies continuously with p and tends to ∞ as $p \to 2$- and $\rho = 1$ when $(1+r)(1-r)^{2/p-1} = 1$, the theorem now follows. Note that if we consider $(f(r) - f(0))/r$ for the extremal f in the theorem and let $r \to 0+$, the result agrees with Theorem 2. ∎

We now consider in detail the special case $L(f) = a_n$. Since $f \in N_p$ if and only if $\gamma f(ze^{i\alpha}) \in N_p$ for all γ and α, $|\gamma| = 1, \alpha$ real, we may assume that for the extremal, $f(0) > 0$ and $a_n > 0$. The following definition will be useful.

Definition 1. If $f(z) = a_0 + a_1 z + \cdots$ and $g(z) = b_0 + b_1 z + \cdots +$ are analytic in a neighborhood of 0, then $f \overset{n}{\approx} g$ means $a_0 = b_0, \; a_1 = b_1, \cdots, a_n = b_n$.

Remark 3. Note that if $f \overset{n}{\approx} g$, h is analytic in a neighborhood of 0 and F is analytic in a neighborhood of a_0, then

$$\begin{aligned} &(i) \quad h \cdot f \overset{n}{\approx} h \cdot g \\ &(ii) \quad F \circ f \overset{n}{\approx} F \circ g. \end{aligned}$$

Using Theorem 1 and assuming $L(f) = a_n$, we see that h is a polynomial of degree $\leq n$ for the extremal $f(z) = (h(z))^{2/p} I(z)$ and that we may take

$$h(0) = c_0 > 0 \quad \text{and} \quad I(z) = e^{-tP(z)}, \quad tP(z) = \sum_{j=1}^{m} t_j \frac{1 + ze^{i\alpha_j}}{1 - ze^{i\alpha_j}},$$

where $0 \leq m \leq n$, each $t_j > 0$ and each factor $(1 - ze^{i\alpha_j})$ is a factor of h. Now set $\hat{h}(z) = z^n \overline{h\left(\frac{1}{\bar{z}}\right)}$ so that $|\hat{h}(e^{i\alpha})|^2 = e^{-in\alpha}[h(e^{i\alpha})\hat{h}(e^{i\alpha})]$ and note that $h\hat{h}$ is a self-inversive polynomial of degree $n+$ degree $h \leq 2n$. Further, each zero of h that lies on $|z| = 1$ is also a zero of $\hat{h}$ and therefore is a zero of $h\hat{h}$ of even multiplicity. Using Theorem 1 (i), for the extremal, we see that

$$a_n + \sum_{k=0}^{n-1} (a_k e^{i(n-k)\alpha} + \overline{a_k} e^{-i(n-k)\alpha}) = a_n e^{in\alpha} h(e^{-i\alpha}) \hat{h}(e^{-i\alpha})$$

and thus,

$$a_0 + \bar{a}_1 z + \bar{a}_2 z^2 + \cdots + a_n z^n + a_{n-1} z^{n+1} + \cdots + a_0 z^{2n} = a_n \overline{h(\bar{z}) \hat{h}(\bar{z})},$$

when $|z| = 1$. This equality therefore persists for all z and we see that $a_0 = a_n c_0 c_n$ so that $c_n > 0$ and degree $h = n$. Further,

$$a_0 + a_1 z + \cdots + a_n z^n + \bar{a}_{n-1} z^{n+1} + \cdots + a_0 z^{2n} = a_n h(z) \hat{h}(z).$$

We now state the next theorem.

Theorem 4. *If $f \in N_p$ satisfies $f(0) > 0$, $a_n > 0$ and $a_n \geq |\frac{g^{(n)}(0)}{n!}|$ for all $g \in N_p$ then*

(i) $f(z) = h^{2/p} I$, where h is a polynomial of degree n, $I = e^{-tP(z)}$, where

$$tP(z) = \sum_{j=1}^{m} t_j \frac{1 + ze^{i\alpha_j}}{1 - ze^{i\alpha_j}}$$

for some m, $0 \leq m \leq n$, each $t_j > 0$, $t = t_1 + \cdots + t_m$ and each $(1 - ze^{i\alpha_j})$ is a factor of h,

(ii) $f \overset{n}{\approx} a_n h\hat{h}$, and

(iii) for $1 \leq k \leq n$,

$$\sum_{j=1}^{m} t_j e^{ik\alpha_j} = \frac{1}{k}\left(1-\frac{1}{p}\right)\sum_{j=1}^{m} e^{ik\alpha_j} + \frac{1}{2k}\sum_{j=m+1}^{n}\left[\frac{1}{\rho_j^k} + \rho_j^k\left(1-\frac{2}{p}\right)\right] e^{ik\alpha_j},$$

where $h(z) = c\prod_{j=1}^{m}(1-ze^{i\alpha_j}z)\prod_{j=m+1}^{n}(1-\rho_j e^{i\alpha_j}z)$, $0 < \rho_j \le 1$ *for each* j, $m < j \le n$.

Proof. (i) and (ii) follow from the discussion preceding the statement of the theorem. To obtain (iii), use the fact that $\frac{1}{2}\log f \overset{n}{\approx} \frac{1}{2}\log(a_n h\hat{h})$. ■

Actually, if the extreme value for $a_n = a_n(p)$ is known to occur when h given in Theorem 4 has all its zeros on $|z| = 1$, $1 < p$, then $a_n(p) = (a_n(\infty))^{1-1/p}$. Clearly, this supports the conjecture that $a_n(p) = (2/e)^{1-1/p}$ since $a_n(\infty)$ is known to be $2/e$, $n = 2, 3, 4$, with the extremal for $p = \infty$ occuring only when $h(z) = \frac{1}{\sqrt{2}}(1+z^n)$ or a rotation. Incidentally, Theorem 4, (ii) can hold for $h(z) = c(1+z)(1+\alpha z\cos\alpha + z^2)$ when $n = 3$ and $h \neq \frac{1}{\sqrt{2}}(1+z^3)$ (see [3]). That is there are are other "local" maxima for a_3 than the one that yields the absolute maximum.

Theorem 5. *If all zeros of h lie on* $|z| = 1$ *and* $f_p(z) = h^{2/p}I$, $1 < p$, *then (ii) of Theorem 4 holds if and only if* $f_\infty(z) \overset{n}{\approx} a_n(\infty)h\hat{h}$ *and* $a_n(p) = (a_n(\infty))^{1-1/p}$.

Proof. Given $h \in H^2$ of degree n with all zeros of h on $|z| = 1$, $h(0) > 0$, $\hat{h}(z) = h(z)$, we have $f_\infty(z) = e^{-tP(z)}$ satisfies $f_\infty(z) \overset{n}{\approx} a_n(\infty)h\hat{h}$ if and only if $e^{-t(1-1/p)P(z)} \overset{n}{\approx} (a_n(\infty))^{1-1/p}h^{2-2/p}$ if and only if $h^{2/p}e^{-t(1-1/p)P(z)} \overset{n}{\approx} (a_n(\infty))^{1-1/p}h\hat{h}$ using Remark 3. The theorem now follows. ■

Further, the inner part of f cannot be trivial for the extremal when $p > 1$. We believe it is trivial when $p < 1$.

Theorem 6. *If* $f \in N_p$ *is extremal for the problem* $L(f) = a_n$, $n > 0$, *and* $1 < p$, *then the inner part I in the representation* $f(z) = (h)^{2/p}I$ *is non-trivial (i.e.* $I \not\equiv 1$*).*

Proof. We know that if I is trivial then $(h)^{2/p} \overset{n}{\approx} a_n h\hat{h}$. Hence, with $h(z) = c_0 + \cdots + c_n z^n$, we have $c_0^{2/p} = a_n c_0 c_n$. However, since $h(z) \neq 0$ in D and $||h||_2 = 1$, $0 < |c_n| \le |c_0| < 1$. Thus, $c_0^{2/p} \le a_n c_0^2$ so that $c_0^{2/p-2} \le a_n$. However $p > 1$ implies $2/p - 2 < 0$ so $a_n > 1$. We know this last inequality is false and it follows that I cannot be trivial. ■

We have a complete solution for $n = 2$.

Theorem 7. *If* $f \in N_p$, *then*

$$|a_2| \le (2/e)^{1-1/p}, \quad 1 \le p, \tag{8}$$

$$|a_2| \le \left(\frac{2}{2-p}\right)^{1-2/p} \cdot \frac{2}{p}, \quad p < 1, \tag{9}$$

with equality if and only if

$$f(z) = \begin{cases} 2^{-1/p}(1+z^2)^{2/p}e^{-(1-1/p)\frac{1-z^2}{1+z^2}}, & 1 \le p, \\ \left(\frac{2-p}{2} + \sqrt{\frac{p(2-p)}{2}}z + \frac{p}{2}z^2\right)^{2/p}, & p < 1, \end{cases} \tag{10}$$

or a rotation.

Proof. Using Theorems 4 and 5, (8) clearly holds when $m = 2$ with equality as stated when $p \geq 1$. Further, $f \in N_p$ implies $(f(z)\overline{f(\bar{z})})^{1/2} \in N_p$. However,

$$\begin{aligned} g(z) &= a_0 \left(1 + 2\mathrm{Re}\frac{a_1}{a_0}z + \left(2\frac{a_2}{a_0} + \frac{|a_1|^2}{2a_0^2}\right)z^2 + \cdots\right)^{1/2} \\ &= a_0 + 2\mathrm{Re}\ a_1 z + \left(a_2 + \frac{|a_1|^2 - (\mathrm{Re}\ a_1)^2}{2a_0}\right)z^2 + \cdots, \end{aligned}$$

so clearly a_1 is real for the extremal and in fact we may assume f has real coefficients. In case $m = 0$, $f(z) = (c_0 + c_1 z + c_2 z^2)^{2/p}$, $c_0 > 0$, $c_2 > 0$, we then have either

$$\text{(11)} \qquad \begin{cases} 0 = \left(\dfrac{1}{\rho} + \rho(1 - \dfrac{2}{p})\right)\cos\alpha, \\ \\ 0 = \left(\dfrac{1}{\rho^2} + \rho^2(1 - \dfrac{2}{p})\right)\cos 2\alpha, \end{cases}$$

or

$$\text{(12)} \qquad \begin{cases} 0 = \dfrac{1}{\rho_1} + \dfrac{1}{\rho_2} + (\rho_1 + \rho_2)(1 - \dfrac{2}{p}), \\ \\ 0 = \dfrac{1}{\rho_1^2} + \dfrac{1}{\rho_2^2} + (\rho_1^2 + \rho_2^2)(1 - \dfrac{2}{p}). \end{cases}$$

In the second case,

$$0 = (\rho_1 + \rho_2)\left(\frac{1}{\rho_1\rho_2} + (1 - \frac{2}{p})\right) = (\rho_1^2 + \rho_2^2)\left(\frac{1}{\rho_1^2\rho_2^2} + (1 - \frac{2}{p})\right).$$

Thus $2/p - 1 = \dfrac{1}{\rho_1^2\rho_2^2} = \dfrac{1}{\rho_1\rho_2}$ so $\rho_1\rho_2 = 1$ and thus $\rho_1 = \rho_2 = \pm 1$, $p = 1$. This case is done.

In the first case, $\cos\alpha \neq 0$ yields $(2/p - 1) = \dfrac{1}{\rho^2}$, $\rho = \sqrt{\dfrac{p}{2-p}}$ and either

$$0 = \frac{2-p}{p} + \frac{p}{2-p}\left(\frac{p-2}{p}\right) = \frac{2-p}{p} - 1 = \frac{2}{p} - 2, \quad p = 1$$

or $\cos 2\alpha = 0$. Thus $\cos\alpha \neq 0$ implies $\cos 2\alpha = 0$. Then $\dfrac{1}{\rho^2} = \dfrac{2-p}{p}$, $\cos\alpha = \pm\dfrac{1}{\sqrt{2}}$, and we may take $\cos\alpha = \frac{1}{\sqrt{2}}$ (i.e. replace $f(z)$ by $f(-z)$ if necessary). Thus,

$$\begin{aligned} f(z) &= c\left[\left(1 + \sqrt{\frac{p}{2-p}}ze^{i\pi/4}\right)\left(1 + \sqrt{\frac{p}{2-p}}ze^{-i\pi/4}\right)\right]^{2/p} \\ &= c\left(1 + \sqrt{2}\sqrt{\frac{p}{2-p}}z + \frac{p}{2-p}z^2\right)^{2/p} \quad \text{where } c = \left(\frac{2-p}{2}\right)^{2/p}, \end{aligned}$$

$$f(z) = \left(\frac{2-p}{2} + \sqrt{\frac{p(2-p)}{2}} + \frac{p}{2} z^2 \right)^{2/p}.$$

Of course this is only a possible extremal for $p \le 1$ and $\frac{1}{\rho^2} = \frac{2-p}{p} \ge 1$. In fact this yields the bound given in the theorem.

If $\cos\alpha = 0$, then $f(z) = g(z^2)$ where $g(z) = [c(1+\rho z)]^{2/p}$ and this implies $p \le 1$ for such an extremal. It is easy to see from Theorem 2 and the above result that $g(z^2)$ cannot be extremal for a_2.

It remains to check $m = 1$. In this case,

$$h(z) = c(1+z)(1+\rho z) = c(1+(1+\rho)z + \rho z^2),$$

where $\frac{1}{c^2} = 2(1+\rho+\rho^2)$ while

$$\begin{aligned} t &= 1 - \frac{1}{p} + \frac{1}{2}\left(\frac{1}{\rho} + \rho\left(1 - \frac{2}{p}\right)\right) \\ &= \frac{1}{2}\left(1 - \frac{1}{p}\right) + \frac{1}{4}\left(\frac{1}{\rho^2} + \rho^2\left(1 - \frac{2}{p}\right)\right). \end{aligned} \tag{13}$$

Therefore,

$$1 - \frac{1}{p} = \frac{(1+\rho)(1-\rho)^3}{2\rho^2(1+2\rho-\rho^2)} > 0 \quad (\text{hence } \; p > 1). \tag{14}$$

When $\rho = 1$, $p = 1$ and p increases as ρ decreases and $p = \infty$ when

$$1 + \frac{1}{2}\left(\rho + \frac{1}{\rho}\right) = \left[\frac{1}{2}\left(\rho + \frac{1}{\rho}\right)\right]^2$$

(using (13)). Thus, p assumes all values from 1 to ∞ as ρ decreases from 1 to $(1 + \sqrt{5} - \sqrt{2+2\sqrt{5}})/2$. Using (13) and (14) together with Theorem 4, an extremal satisfies $c_0^{2/p} e^{-t} = a_2 c_0 c_2$, $c_0 = (2(1+\rho+\rho^2))^{-1/2}$, $c_2 = \rho c_0$ and hence

$$a_2 = \frac{1}{\rho}[2(1+\rho+\rho^2)]^{1-1/p} e^{-\left[(1-1/p)(1+\rho) + \frac{1}{2}\frac{1-\rho^2}{\rho}\right]}.$$

Thus, to complete the proof of the theorem, we require

$$\frac{1}{\rho}(1+\rho+\rho^2)^{1-1/p} e^{-\left[(1-1/p)\rho + \frac{1}{2}\frac{1-\rho^2}{\rho}\right]} \le 1.$$

On taking logs and using (14) we require

$$\frac{(1+\rho)(1-\rho)^3}{2\rho^2(1+2\rho-\rho^2)} \left[\log(1+\rho+\rho^2) - \rho\right] \le \frac{1}{2}\frac{1-\rho^2}{\rho} + \log\rho.$$

Now

$$\log \rho = \log(1-(1-\rho)) \geq -(1-\rho) - \frac{(1-\rho)^2}{2} - \frac{(1-\rho)^3}{3}\sum_{k=0}^{\infty}(1-\rho)^k$$

$$= -(1-\rho) - \frac{(1-\rho)^2}{2} - \frac{(1-\rho)^3}{3\rho}.$$

Further $\dfrac{1+\rho}{1+2\rho-\rho^2} = \dfrac{1+\rho}{1+\rho+(\rho-\rho^2)} \leq 1$. Thus, it is sufficient to show

$$(1-\rho)^3 \log(1+\rho+\rho^2) \leq \rho\left[1-\rho^2-2\rho(1-\rho)-\rho(1-\rho)^2-\tfrac{2}{3}(1-\rho)^3+(1-\rho)^3\right]$$

$$= \frac{4(1-\rho)^3}{3}\rho.$$

Finally, $1+\rho+\rho^2 \leq e^{\frac{4}{3}\rho} = 1+\frac{4}{3}\rho+\frac{8}{9}\rho^2+\cdots$ is sufficient. The last inequality clearly holds since $0 < \frac{1}{3}\rho - \frac{1}{9}\rho^2, \quad 0 < \rho \leq 1$. This completes the proof. ■

References

[1] J. E. Brown, *On a Coefficient Problem for nonvanishing H_p functions*, Complex Variables, Theory and Applications **4** (1985), 253-265.

[2] C. Carathéodory, *Über den Variabilitätsbereich der Fourierschen Konstanten von positiven harmonischen Funktionen*, Rend. Circ. Mat. Palermo **32** (1911), 193-217.

[3] J. A. Hummel, S. Scheinberg and L. Zalcman, *A coefficient problem for bounded nonvanishing functions*, J. Analyse Math. **31** (1977), 169-190.

[4] J. Krzyż, *Coefficient problems for bounded nonvanishing functions*, Ann. Polon. Math. **20** (1968), 314.

[5] Delin Tan, *Coefficient estimates for bounded nonvanishing functions*, Chinese Ann. Math. Ser. **A4** (1983), 97-104 (Chinese).

[6] O. Toeplitz, *Über die Fouriersche Entwicklung positiver Funktionen*, Rend. Circ. Math. Palermo **32** (1911), 191-192.

Received: September 3, 1989.

Computational Methods and Function Theory
Proceedings, Valparaíso 1989
St. Ruscheweyh, E.B. Saff, L. C. Salinas, R.S. Varga (*eds.*)
Lecture Notes in Mathematics **1435**, pp. 191–200

Some results on separate convergence of continued fractions

W.J. Thron[1]

Department of Mathematics, University of Colorado
Campus Box 426, Boulder, CO 80309, U.S.A.

1. Introduction

The term *separate convergence* was introduced recently to describe the phenomena which result when conditions — stronger than needed for convergence — are imposed on continued fractions or similar algorithms. The term *separate* is motivated by the first result of this type due to Śleszyński in 1888. He proved that if the continued fraction $K(a_n z/1)$ satisfies the condition $\sum |a_n| < \infty$ then not only does the sequence of approximants $\{A_n(z)/B_n(z)\}$ converge but $\{A_n(z)\}$ and $\{B_n(z)\}$ converge separately to entire functions $A(z)$ and $B(z)$.

In later investigation the conclusions frequently are less sweeping. One settles for the existence of an "easily described" sequence $\{\Gamma_n(z)\}$ such that $\{A_n(z)/\Gamma_n(z)\}$ and $\{B_n(z)/\Gamma_n(z)\}$ converge separately for $z \in \Delta$. Usually the convergence will be uniform on compact subsets of Δ, which may be a proper subset of the complex plane $\mathbb{C}$. The restriction that the Γ_n can be "easily described" is essential because for convergent continued fractions one can always choose $\Gamma_n(z) = B_n(z)$ and thus the distinction between ordinary and separate convergence would become meaningless.

Most, if not all, instances of separate convergence occur for limit periodic continued fractions with elements that are functions of a complex variable. Sometimes separate convergence is a tool in the derivation of results on analytic continuation or behavior on the boundary of the function to which the continued fraction converges. In other cases it may only be a by-product of such investigations. Since orthogonal polynomials can be obtained as denominators of the approximants of certain continued fractions, there is an overlap between our work and known results about the asymptotic behavior of

[1]This research was supported in part by the U.S. National Science Foundation under grant No. DMS-8700498.

orthogonal polynomials (see, for example [2]). Our results can also be used to obtain information on the asymptotic behavior of orthogonal L–polynomials.

Instead of describing at length the results obtained on separate convergence, we give in the References a list of all articles on the subject we know about. This includes some papers where separate convergence is provable, but is not explicitly established.

We shall concentrate here on the proof of a basic theorem from which some, but not all, of the known results on separate convergence can be derived. The theorem also yields new results, in particular on general T-fractions. We shall also discuss briefly the impact of equivalence transformations on separate convergence.

An account concerned mainly with separate convergence of PC fractions, Schur fractions and Schur algorithms is being prepared by O. Njåstad.

In a subsequent article we hope to return to the subject and explore some other general approaches which yield results on separate convergence.

2. Asymptotic behavior of numerators and denominators of approximants of limit periodic continued fractions.

In this section we shall prove a quite general and rather involved theorem. In the next section corollaries of a much simpler nature will be given.

Theorem 2.1. *Let $a_n(z)$, $b_n(z)$, $a(z)$ and $b(z)$ be holomorphic functions for $z \in \Delta$. Further assume that $a_n(z) \neq 0$ and*

$$\lim_{n\to\infty} a_n(z) = a(z), \quad \lim_{n\to\infty} b_n(z) = b(z) \quad \text{for} \quad z \in \Delta.$$

Then

$$\operatorname*{K}_{n=1}^{\infty} \left(\frac{a_n(z)}{b_n(z)}\right) \tag{2.1}$$

is a limit periodic continued fraction for $z \in \Delta$. Set

$$a_n(z) = a(z) + \delta_n(z), \; b_n(z) = b(z) + \eta_n(z). \tag{2.2}$$

Note that $\delta_n(z)$ and $\eta_n(z)$ are holomorphic and

$$\lim_{n\to\infty} \delta_n(z) = 0, \quad \lim \eta_n(z) = 0$$

for $z \in \Delta$. Let $x_1(z)$ and $x_2(z)$ be the solutions of

$$w^2 + b(z)w - a(z) = 0$$

and assume that the solutions have been so numbered that

$$\left|\frac{x_1(z)}{x_2(z)}\right| < 1 \quad \text{for} \quad z \in \Delta^* \subset \Delta. \tag{2.3}$$

Further assume that the series

$$\sum_{k=1}^{\infty} |\delta_k(z)| \quad \text{and} \quad \sum_{k=1}^{\infty} |\eta_k(z)| \tag{2.4}$$

converge uniformly on compact subsets of $\Delta_0 \subset \Delta$ *and that*

$$|x_2(z)| > 0 \quad \textit{for} \quad z \in \Delta_0. \tag{2.5}$$

Finally, let $\Delta^{(\varepsilon)}$ *be such that*

$$|x_1(z) - x_2(z)| > 2\varepsilon \quad \textit{for} \quad z \in \Delta^{(\varepsilon)}. \tag{2.6}$$

Then

$$\lim_{n\to\infty} \frac{A_n(z)}{(-x_2(z))^{n+1}} \quad \text{and} \quad \lim_{n\to\infty} \frac{B_n(z)}{(-x_2(z))^{n+1}} \tag{2.7}$$

both exist and are holomorphic in

$$\Delta^{\dagger} = \bigcup_{\varepsilon>0} \Delta^* \cap \Delta_0 \cap \Delta^{(\varepsilon)}.$$

Here $A_n(z)$ *and* $B_n(z)$ *are the numerator and denominator, respectively, of the nth approximant of the continued fraction (2.1). Of course the theorem is of interest only if* $\Delta^{\dagger} \neq \emptyset$.

Proof. From now on we shall usually not indicate the dependence of the various functions under consideration on z. We shall simply write

$$a_n,\ b_n,\ a,\ b,\ \delta_n,\ \eta_n,\ x_1,\ x_2,\ A_n,\ B_n,\ C_n,\ D_n.$$

We note that

$$-a = x_1 x_2, \qquad -b = x_1 + x_2. \tag{2.8}$$

As was shown in [16] one then has

$$\begin{aligned} B_n + x_1 B_{n-1} &= (b + x_1)B_{n-1} + aB_{n-2} + \eta_n B_{n-1} + \delta_n B_{n-2} \\ &= (-x_2)B_{n-1} + (-x_1x_2)B_{n-2} + \eta_n B_{n-1} + \delta_n B_{n-2}. \end{aligned}$$

Iterating these equations one arrives at

$$B_n + x_1 B_{n-1} = (-x_2)^n + \sum_{k=0}^{n-1} (-x_2)^{n-1-k} \eta_{k+1} B_k + \sum_{k=0}^{n-1} (-x_2)^{n-1-k} \delta_{k+1} B_{k-1}. \tag{2.9}$$

An analogous derivation leads to

$$B_n + x_2 B_{n-1} = (-x_1)^n + \sum_{k=0}^{n-1} (-x_1)^{n-1-k} \eta_{k+1} B_k + \sum_{k=0}^{n-1} (-x_1)^{n-1-k} \delta_{k+1} B_{k-1}. \tag{2.10}$$

If $x_1 \neq x_2$, one can solve the system of equations (2.9), (2.10) for B_n. The result is

$$(x_1 - x_2)B_n = (-x_2)^{n+1} - (-x_1)^{n+1} + \sum_{k=0}^{n-1}((-x_2)^{n-k} - (-x_1)^{n-k})\eta_{k+1}B_k$$

$$+ \sum_{k=0}^{n-1}((-x_2)^{n-k} - (-x_1)^{n-k})\delta_{k+1}B_{k-1}.$$

Introduce

$$D_n = \frac{B_n}{(-x_2)^{n+1}}. \tag{2.11}$$

In terms of D_n we then have

$$\begin{aligned}(x_1 - x_2)D_n = 1 &- \left(\frac{x_1}{x_2}\right)^{n+1} + \sum_{k=0}^{n-1}\left(1 - \left(\frac{x_1}{x_2}\right)^{n-k}\right)\eta_{k+1}D_k \\ &+ \frac{1}{-x_2}\sum_{k=0}^{n-1}\left(1 - \left(\frac{x_1}{x_2}\right)^{n-k}\right)\delta_{k+1}D_{k-1}.\end{aligned} \tag{2.12}$$

Similarly one obtains for

$$C_n = \frac{A_n}{(-x_2)^{n+1}} \tag{2.13}$$

the formula

$$\begin{aligned}(x_1 - x_2)C_n = x_1 &- \left(\frac{x_1}{x_2}\right)^{n+1} x_2 + \sum_{k=0}^{n-1}\left(1 - \left(\frac{x_1}{x_2}\right)^{n-k}\right)\eta_{k+1}C_k \\ &+ \frac{1}{-x_2}\sum_{k=0}^{n-1}\left(1 - \left(\frac{x_1}{x_2}\right)^{n-k}\right)\delta_{k+1}C_{k-1}.\end{aligned} \tag{2.14}$$

From (2.12) one can deduce for $z \in \Delta^*$ the inequality

$$|x_1 - x_2|\,|D_n| \le 2 + 2\sum_{k=0}^{n-1}|\eta_{k+1}|\,|D_k| + \frac{2}{|x_2|}\sum_{k=0}^{n-1}|\delta_{k+1}|\,|D_{k-1}|. \tag{2.15}$$

We would like to prove that there exists a constant $M > 0$ such that for z in certain subsets of $\Delta_0 \cap \Delta^*$

$$|D_n| < M \quad \text{for all} \quad n \ge 1.$$

To do this we prove the following lemma. A similar result can be found in [2, p. 455].

Lemma 2.2. *Let $a > 0$, $\gamma_n \ge 0$, $n \ge 1$ and $S_n \ge 0$, $n \ge 0$. If*

$$S_n \le a + \sum_{k=0}^{n-1}\gamma_{k+1}S_k, \quad n \ge 1,\ S_0 \le a, \tag{2.16}$$

then

$$S_n \le a \prod_{k=1}^{n}(1+\gamma_k) =: P_n, \quad n \ge 1. \tag{2.17}$$

Proof. We have

$$S_1 \le a + \gamma_1 S_0 \le a(1+\gamma_1).$$

Set $P_0 = a$ and assume that $S_k \le P_k$ for $0 \le k \le n-1$. Then

$$\begin{aligned} S_n &\le a + \textstyle\sum_{k=0}^{n-1}((1+\gamma_k)-1)P_k = a + \sum_{k=0}^{n-1}(P_{k+1}-P_k) \\ &= a - P_0 + P_n = P_n. \end{aligned}$$

The lemma is thus proved by induction. ∎

We apply the lemma to (2.15) by setting $|D_n| = S_n$, $a = \max(2/|x_1-x_2|,\ 1/|x_2|)$,

$$\gamma_k = \frac{2}{|x_1-x_2|}\left(|\eta_k| + |\frac{\delta_{k+1}}{x_2}|\right).$$

For K a compact subset of $\Delta_0 \cap \Delta^*$ we then have $|x_2| > d_K > 0$, $|x_1/x_2| < r_K < 1$. In view of (2.4) the sequence $\{D_n\}$ has a uniform bound $M(K,\varepsilon)$ satisfying

$$|D_n| \le \max(1/\varepsilon, 1/d_K)\prod_{k=1}^{\infty}\left(1+|\eta_k|+\frac{|\delta_{k+1}|}{d_K}\right) \le M(K,\varepsilon). \tag{2.18}$$

for $z \in K \cap \Delta^{(\varepsilon)}$.

Let $z \in K \cap \Delta^{(\varepsilon)}$, set $|x_1/x_2| = r \le r_K < 1$. From (2.12) one deduces

$$\begin{aligned} |x_1-x_2||D_{n+m}-D_n| &\le r^{n+m} + r^n + \sum_{k=0}^{n-1}(r^{n+m-k}+r^{n-k})|\eta_{k+1}||D_k| \\ &+\frac{1}{|x_2|}\sum_{k=0}^{n-1}(r^{n+m-k}+r^{n-k})|\delta_{k+1}||D_{k-1}| \\ &+\sum_{k=n}^{n+m-1}(1+r^{n+m-k})|\eta_{k+1}||D_k| \\ &+\frac{1}{|x_2|}\sum_{k=n}^{n+m-1}(1+r^{n+m-k})|\delta_{k+1}||D_{k-1}| \\ &\le 2r_K^n + 2Mr_K^{n/2}\sum_{k=0}^{[n/2]}\left(|\eta_{k+1}|+\frac{|\delta_{k+2}|}{d_K}\right) \\ &+2M\sum_{k=[n/2]}^{\infty}\left(|\eta_k|+\frac{|\delta_{k+1}|}{d_K}\right). \end{aligned} \tag{2.19}$$

One thus can write for $z \in K \cap \Delta^{(\varepsilon)}$, where K is a compact subset of $\Delta_0 \cap \Delta^*$,

(2.20)
$$|D_{n+m} - D_n| < \frac{1}{\varepsilon}\left(r_K^{n/2} C_1(K,\varepsilon) + \sum_{k=[n/2]}^{\infty}\left(|\eta_k| + \frac{|\delta_{k+1}|}{d_K}\right)\right).$$

It follows that $\{D_n\}$ converges uniformly on compact subsets of $\Delta(\varepsilon) := \Delta_0 \cap \Delta^* \cap \Delta^{(\varepsilon)}$, since every compact subset of $\Delta(\varepsilon)$ is a compact subset of $\Delta_0 \cap \Delta^*$. Hence $D = \lim_{n\to\infty} D_n$ is a holomorphic function on $\Delta(\varepsilon)$. Analytic continuation can then be invoked to conclude that $D(z)$ is holomorphic for all $z \in \Delta^\dagger = \bigcup_{\varepsilon>0} \Delta(\varepsilon)$.

Similarly one shows that
$$\lim \frac{A_n}{(-x_2)^{n+1}} = C$$
exists and is holomorphic for all $z \in \Delta^\dagger$. ∎

3. Applications of Theorem 2.1.

We begin with the theorem of Śleszyński of 1888. Śleszyński's approach also allowed one to conclude that A and B are of order at most one. That A and B have no common zero was first proved by Maillet [4], (see also Perron [5, p. 150]). By further refining Śleszyński's method one can get stronger results on the order and type of the two entire functions A and B. We plan to present results of this kind in a future article [15].

Corollary 3.1. *If the regular C–fraction*

(3.1)
$$\mathop{K}_{n=1}^{\infty}\left(\frac{a_n z}{1}\right)$$

satisfies
$$\sum_{n=1}^{\infty} |a_n| < \infty,$$
then
$$\lim_{n\to\infty} A_n(z) = A(z) \quad \text{and} \quad \lim_{n\to\infty} B_n(z) = B(z)$$
for all $z \in \mathbb{C}$. $A(z)$ and $B(z)$ are entire functions.

Proof. This is the special case of Theorem 2.1 where $a = 0$, $b = 1$, $\delta_n = 0$, $x_1 = 0$, $x_2 = -1$ and hence
$$\Delta = \Delta^* = \Delta_0 = \mathbb{C}, \quad \Delta^{(\varepsilon)} = \mathbb{C}, \quad 0 < \varepsilon < 1.$$ ∎

Corollary 3.2. *If the regular C–fraction (3.1) satisfies*

(3.2)
$$\sum_{n=1}^{\infty} |a_n - a| < \infty, \quad a \neq 0,$$

then

$$\lim_{n\to\infty} \frac{2^{n+1}A_n(z)}{(1+\sqrt{1+4az}\,)^{n+1}} = C(z), \quad \lim_{n\to\infty} \frac{2^{n+1}B_n(z)}{(1+\sqrt{1+4az}\,)^{n+1}} = D(z)$$

for all $z \in \psi_a = \left[z : az \in \mathbb{C} \sim [s : s \in \mathbb{R}, -\frac{1}{4} \le s < -\infty]\right]$. *Here* $\sqrt{\ }$ *is chosen so that* $Re\sqrt{\ } > 0$ *for* $z \in \psi_a$.

Proof. In this case $\Delta = \mathbb{C}$, $\Delta^* = \Delta_0 = \psi_2$ and

$$\Delta^{(\varepsilon)} = \left[z : \left|-\frac{1}{4a} - z\right| > \varepsilon^2/4|a|\right]. \qquad \blacksquare$$

The proof of the convergence of (3.1) under the assumption (3.2) is also due to Śleszyński [13]. Our conclusion about separate convergence is not explicitly stated in his article but can easily be deduced.

Corollary 3.3 *If the general T–fraction*

$$\underset{n=1}{\overset{\infty}{K}} \left(\frac{F_n z}{1+G_n z}\right) \tag{3.3}$$

satisfies $\sum_{n=1}^{\infty} |F_n| < \infty$ *and* $\sum_{n=1}^{\infty} |G_n - G| < \infty$, $G \neq 0$, *then*

$$\lim_{n\to\infty} \frac{A_n(z)}{(1+Gz)^{n+1}} = C^{\dagger}(z), \quad \lim \frac{B_n(z)}{(1+Gz)^{n+1}} = D^{\dagger}(z)$$

for all $z \in \mathbb{C} \sim [-1/G]$. *The functions* $C^{\dagger}$ *and* $D^{\dagger}$ *are holomorphic in* $\mathbb{C} \sim [-1/G]$.

Proof. $x_1 = 0$, $x_2 = -1 - Gz$, $\Delta = \mathbb{C}$, $\Delta^* = \mathbb{C} \sim [-1/G] = \Delta_0$, $\Delta^{(\varepsilon)} = [z : |1+Gz| > 2\varepsilon]$. $\blacksquare$

Corollary 3.4. *In the general T–fraction (3.3) let*

$$\lim_{n\to\infty} F_n = 1, \qquad \lim_{n\to\infty} G_n = -1. \tag{3.4}$$

Further, assume that

$$\sum_{n=1}^{\infty} |F_n - 1| < \infty, \qquad \sum_{n=1}^{\infty} |G_n + 1| < \infty.$$

Then

$$\lim_{n\to\infty} A_n(z) = A(z), \qquad \lim B_n(z) = B(z)$$

and A *and* B *are holomorphic for* $|z| < 1$. *Also*

$$\lim_{n\to\infty} \frac{A_n(z)}{z^{n+1}} = \hat{C}(z), \quad \lim_{n\to\infty} \frac{B_n(z)}{z^{n+1}} = \hat{D}(z)$$

and $\hat{C}$ *and* $\hat{D}$ *are holomorphic for* $|z| > 1$.

Proof. $\Delta = \mathbb{C}$. 1st case: $x_1 = -z$, $x_2 = -1$, $\Delta^* = [z : |z| < 1]$, $\Delta^{(\varepsilon)} = [z : |z-1| > 2\varepsilon]$. 2nd case: $x_1 = -1$, $x_2 = -z$, $\Delta^* = [z : |z| > 1]$, $\Delta^{(\varepsilon)} = [z : |1-z| > 2\varepsilon]$. $\blacksquare$

This corollary overlaps some results of Waadeland [21] and Thron and Waadeland [15].

Corollary 3.5. *In the J–fraction*

$$\underset{n=1}{\overset{\infty}{K}} \left(\frac{k_n}{z - c_n} \right) \tag{3.5}$$

let $\sum_{n=1}^{\infty} |k_n| < \infty$, $\sum_{n=1}^{\infty} |c_n - c| < \infty$. *Then*

$$\lim_{n\to\infty} \frac{A_n(z)}{(c-z)^{n+1}} = C^*(z), \quad \lim_{n\to\infty} \frac{B_n(z)}{(c-z)^{n+1}} = D^*(z)$$

and C^* *and* D^* *are holomorphic in* $\mathbb{C} \sim [c]$.

Proof. $\Delta = \Delta^* = \mathbb{C}$, $x_1 = 0$, $x_2 = z - c$, $\Delta_0 = \mathbb{C} \sim [c]$, $\Delta^{(\varepsilon)} = [z : |z - c| > 2\varepsilon]$. This is Theorem 5 in Schwartz [12]. ∎

4. Changes in separate convergence under equivalence transformations.

Let $K_{n=1}^{\infty} (a_n/b_n)$ be a given continued fraction, let

$$\underset{n=1}{\overset{\infty}{K}} \left(\frac{\hat{a}_n}{\hat{b}_n} \right) \tag{4.1}$$

be defined by

$$\hat{a}_n = \gamma_{n-1}\gamma_n a_n, \quad \hat{b}_n = \gamma_n b_n, \quad \gamma_0 = 1, \quad \gamma_n \neq 0, \quad n \geq 1. \tag{4.2}$$

Then the two continued fractions are equivalent in the sense that their sequences of approximants are identical. However for the numerators and denominators of the approximants one can show, using the recursion relations, that

$$\hat{A}_n = A_n \prod_{v=0}^{n} \gamma_v, \quad \hat{B}_n = B_n \prod_{v=0}^{n} \gamma_v, \quad n \geq -1, \tag{4.3}$$

provided one uses the convention $\prod_k^m = 1$, for $m < k$.

We use this idea to prove another result for J–fractions (3.4). The continued fraction

$$K \left(\frac{k_n (z - c_n)^{-1} (z - c_{n-1})^{-1}}{1} \right), \quad c_0 = z - 1, \tag{4.4}$$

is equivalent to (3.4). Under the conditions

$$c_n \to \infty, \quad \sum_{n=1}^{\infty} \left| \frac{k_n}{c_n c_{n-1}} \right| < \infty, \tag{4.5}$$

the numerators $\hat{A}_n$ and denominators $\hat{B}_n$ of (4.4) satisfy, for $z \in \mathbb{C} \sim [c_1, c_2 \cdots]$

$$\lim_{n\to\infty} \hat{A}_n = A, \qquad \lim \hat{B}_n = B.$$

It follows from (4.3) that for the J–fraction (3.4) one has

$$\lim_{n\to\infty} \frac{A_n}{\prod_{v=0}^{n}(z-c_v)} = A, \qquad \lim_{n\to\infty} \frac{B_n}{\prod_{v=0}^{n}(z-c_v)} = B.$$

If one strengthens (4.5) to

$$\sum_{n=1}^{\infty} \left|\frac{1}{c_n}\right| < \infty, \qquad \sum_{n=1}^{\infty} \left|\frac{k_n}{c_n c_{n-1}}\right| < \infty, \tag{4.6}$$

one obtains (Theorem 2 in Schwartz [10])

$$\lim_{n\to\infty} \frac{A_n}{\prod_{v=1}^{n}(-c_v)} = A \prod_{v=1}^{\infty}\left(1-\frac{z}{c_v}\right), \quad \lim \frac{B_n}{\prod_{v=1}^{n}(-c_v)} = B \prod_{v=1}^{\infty}\left(1-\frac{z}{c_v}\right).$$

References

[1] R. J. Arms, A. Edrei, *The Padé tables and continued fractions generated by totally positive sequences*, Mathematical Essays dedicated to A. J. Macintyre, Athens, Ohio (1970), 1-21.

[2] F. W. Atkinson, *Discrete and continuous boundary problems*, Academic Press, New York, 1964.

[3] A. Auric, *Recherches sur les fractions continues algébriques*, J. Math pure et applique, (6) **3** (1907), 105-206.

[4] A. Edrei, *Sur des suites de nombres liées à la théorie des fractions continues*, Bull. Sci. Math., (2) **72** (1948), 45-64.

[5] Lisa Jacobsen, *A note on separate convergence of continued fractions*, preprint.

[6.] E. Maillet, *Sur les fractions continues algébriques*, J. Ec. pol., (2) **12** (1908), 41-62.

[7] O. Perron, *Die Lehre von den Kettenbrüchen*, 3. Aufl., 2. Band, Teubner, Stuttgart, 1957.

[8.] H.-J. Runckel, *Bounded analytic functions in the unit disk and the behavior of certain analytic continued fractions near the singular line*, J. reine angew. Math., **281** (1976), 97-125.

[9] ___________, *Continuity on the boundary and analytic continuation of continued fractions*, Math. Zeitschr., **148** (1976), 189-205.

[10] ___________, *Meromorphic extension of analytic continued fractions*, Rocky Mtn. J. Math., to appear.

[11] J. Schur, *Über Potenzreihen die im Innern des Einheitskreises beschränkt sind (Fortsetzung)*, J. reine angew. Math., **148** (1918/19), 122-145.

[12] H. M. Schwartz, *A class of continued fractions*, Duke Math. J., **6** (1940), 48-65.

[13] I. V. Sleshinskii (J. Śleszyński), *Convergence of continued fractions* (in Russian), Zapiski matematicheskago otodieleniia Novorossiiskago obshchestvoispytatela, **8** (1888), 97-127.

[14] T. J. Stieltjes, *Recherches sur les fractions continues*, Ann. Fac. Sci. Toulouse, **8** (1894), 1-122.

[15] W. J. Thron, *Order and type of entire functions arising from separately convergent continued fractions*, submitted.

[16] W. J. Thron and H. Waadeland, *Convergence questions for limit periodic continued fractions*, Rocky Mtn. J. Math., **11** (1981), 641-657.

[17] Walter van Assche, *Asymptotics for orthogonal polynomials and three term recurrences*, Proceedings of Columbus conference, to appear.

[18] H. von Koch, *Quelques théorèmes concernant la theorie générale des fractions continues*, Översigt av kongl. Vetenskaps–Akademiens Förhandlingar, **52** (1895).

[19] ___________, *Sur la convergence des determinants d'ordre infini et des fractions continues*, Comptes Rendus, **120** (1895), 145.

[20] ___________, *Sur un théorème de Stieltjes et sur les fonctions définies par des fractions continues*, Bull. Soc. Math. France, **23** (1895), 33-41.

[21] H. Waadeland, *A convergence property of certain T–fraction expansions*, Kgl. norske videnskabers selskabs skrifter 1966, No. 9, 3-22.

[22] H. S. Wall, *Some recent developments in the theory of continued fractions*, Bull. Amer. Math. Soc., **47** (1941), 405-423.

[23] ___________, *The behavior of certain Stieltjes continued fractions near the singular line*, Bull. Amer. Math. Soc., **48** (1942), 427-431.

Received: August 7, 1989, in revised form November 28, 1989.

Computational Methods and Function Theory
Proceedings, Valparaíso 1989
St. Ruscheweyh, E.B. Saff, L. C. Salinas, R.S. Varga (*eds.*)
Lecture Notes in Mathematics **1435**, pp. 201–207

Asymptotics for the Zeros of the Partial Sums of e^z. II

R.S. Varga*

Institute for Computational Mathematics
Kent State University, Kent, OH 44242, USA

and

A.J. Carpenter

Department of Mathematical Sciences
Butler University, Indianapolis, IN 46208, USA

1. Introduction

With $s_n(z) := \sum_{j=0}^{n} z^j/j!$ $(n = 1, 2, \cdots)$ denoting the familiar partial sums of the exponential function e^z, we continue our investigation here on the location of the zeros of the *normalized* partial sums, $s_n(nz)$, which are known to lie (cf. Anderson, Saff, and Varga [1]) for every $n > 1$ in the open unit disk $\Delta := \{z \in \mathbb{C} : |z| < 1\}$. For notation, let the Szegö curve, D_∞, be defined by

$$D_\infty := \{z \in \mathbb{C} : |ze^{1-z}| = 1 \text{ and } |z| \le 1\}. \tag{1.1}$$

It is known that D_∞ is a simple closed curve in the closed unit disk $\bar{\Delta}$, and that D_∞ is star-shaped with respect to the origin, $z = 0$.

If $\{z_{k,n}\}_{k=1}^{n}$ denotes the zeros of $s_n(nz)$ (for $n = 1, 2, \cdots$), then it was shown by Szegö [7] in 1924 that each accumulation point of all these zeros, $\{z_{k,n}\}_{k=1,n=1}^{n,\infty}$, must lie on D_∞, and, conversely, that each point of D_∞ is an accumulation point of the zeros $\{z_{k,n}\}_{k=1,n=1}^{n,\infty}$. Subsequently, it was shown by Buckholtz [2] that the zeros $\{z_{k,n}\}_{k=1,n=1}^{n,\infty}$ all lie *outside* the simple closed curve D_∞.

As for a measure of the rate at which the zeros, $\{z_{k,n}\}_{k=1}^{n}$, tend to D_∞, we use the quantity

*Research supported by the National Science Foundation

$$\text{dist}\,[\{z_{k,n}\}_{k=1}^{n}; D_\infty] := \max_{1\le k\le n}(\text{dist}\,[z_{k,n}; D_\infty]), \tag{1.2}$$

and a result of Buckholtz [2] gives that

$$\text{dist}\,[\{z_{k,n}\}_{k=1}^{n}; D_\infty] \le \frac{2e}{\sqrt{n}} = \frac{5.43656\cdots}{\sqrt{n}} \qquad (n = 1, 2, \cdots) \tag{1.3}$$

It was later shown in Carpenter, Varga, and Waldvogel [3] that the result of (1.3) is *best possible*, as a function of n, since

$$\varliminf_{n\to\infty}\{\sqrt{n}\cdot \text{dist}\,[\{z_{k,n}\}_{k=1}^{n}; D_\infty]\} \ge 0.63665\cdots > 0. \tag{1.4}$$

It was also shown in [3] that there is substantially *faster* convergence of the subset of the zeros $\{z_{k,n}\}_{k=1}^{n}$, to the Szegö curve D_∞, which stay *uniformly* away from the point $z = 1$. More precisely, for the open disk C_δ about the point $z = 1$, defined by

$$C_\delta := \{z \in \mathbb{C} : |z-1| < \delta\} \qquad (0 < \delta \le 1), \tag{1.5}$$

it was shown in [3] that, for any fixed δ with $0 < \delta \le 1$,

$$\text{dist}\,[\{z_{k,n}\}_{k=1}^{n}\backslash C_\delta; D_\infty] = O\left(\frac{\log n}{n}\right) \qquad (n \to \infty), \tag{1.6}$$

and, the result of (1.6) is also *best possible*, as a function of n, since (cf. [3, eq. (2.27)])

$$\varliminf_{n\to\infty}\left\{\frac{n}{\log n}\cdot \text{dist}\,[\{z_{k,n}\}_{k=1}^{n}\backslash C_\delta; D_\infty]\right\} \ge 0.10890\cdots > 0, \tag{1.7}$$

for any fixed δ with $0 < \delta \le 1$.

In [3], an arc, D_n, was defined for each $n = 1, 2, \cdots$ by

$$D_n := \left\{z \in \mathbb{C} : |ze^{1-z}|^n = \tau_n\sqrt{2\pi n}\left|\frac{1-z}{z}\right|,\ |z| \le 1, \text{ and } |\arg z| \ge \cos^{-1}\left(\frac{n-2}{n}\right)\right\}, \tag{1.8}$$

where from Stirling's formula,

$$\tau_n := \frac{n!}{n^n e^{-n}\sqrt{2\pi n}} \approx 1 + \frac{1}{12n} + \frac{1}{288n^2} - \frac{139}{51840n^3} + \cdots, \text{ as } n \to \infty. \tag{1.9}$$

This arc was introduced to provide a much closer approximation to the zeros $\{z_{k,n}\}_{k=1}^{n}$ of $s_n(nz)$, than does the Szegö curve. With the notation of (1.5), it was shown in [3] that, for any fixed δ with $0 < \delta \le 1$,

$$\text{dist}\,[\{z_{k,n}\}_{k=1}^{n}\backslash C_\delta; D_n] = O\left(\frac{1}{n^2}\right) \qquad (n \to \infty), \tag{1.10}$$

and moreover that (1.10) is *best possible*, as a function of n, since (cf. [3, eq. (3.18)])

$$\varliminf_{n\to\infty}\{n^2\cdot \text{dist}\,[\{z_{k,n}\}_{k=1}^{n}\backslash C_\delta; D_n]\} \ge 0.13326\cdots > 0, \tag{1.11}$$

for any fixed δ with $0 < \delta \leq 1$.

It turns out (cf. [3, Prop. 3]) that, for each positive integer n, the arc D_n is *star-shaped* with respect to the origin, $z = 0$, i.e., for each real number θ in $[-\pi, +\pi]$ with $|\theta| \geq \cos^{-1}(\frac{n-2}{n})$, there is a unique positive number $r = r_n(\theta)$ such that $z = re^{i\theta}$ lies on the arc D_n of (1.8). Let $\mathcal{D}_n$ be the closed star-shaped (with respect to $z = 0$) set defined from the arc D_n, i.e.,

$$\mathcal{D}_n := \Big\{ z \in \mathbb{C} : |ze^{1-z}|^n \leq \tau_n \sqrt{2\pi n} \left| \frac{1-z}{z} \right|, |z| \leq 1, \text{ and } |\arg z| \geq \cos^{-1}(\frac{n-2}{n}) \Big\}, \quad (n = 1, 2, \cdots). \tag{1.12}$$

Recently, R. Barnard and K. Pierce asked if the zeros, $\{z_{k,n}\}_{k=1}^{n}$, of $s_n(nz)$ all lie outside $\mathcal{D}_n$ for *every* $n \geq 1$. (This would be the natural analogue of the result of Buckholtz [2] which established that all the zeros $\{z_{k,n}\}_{k=1,n=1}^{n,\infty}$ lie outside of D_∞.) This is not at all obvious from the graphs of [3], since it appeared that the zeros $\{z_{k,16}\}_{k=1}^{16}$ and $\{z_{k,27}\}_{k=1}^{27}$ of $s_{16}(16z)$ and $s_{27}(27z)$ were, to plotting accuracy, respectively *on* the curves D_{16} and D_{27}.

It turns out that the zeros, $\{z_{k,n}\}_{k=1}^{n}$ of $s_n(nz)$ do *not* all lie outside $\mathcal{D}_n$ for every $n \geq 1$. This follows from our first result below (to be established in §2).

Proposition 1 *If $\{z_{k,n}\}_{k=1}^{n}$ denotes the zeros of $s_n(nz)$ with increasing arguments, i.e.,*

$$0 < \arg z_{1,n} \leq \arg z_{2,n} \leq \cdots \leq \arg z_{n,n} < 2\pi, \tag{1.13}$$

then (cf. (1.12)) *$z_{1,n}$ is an element of $\mathcal{D}_n$ for all positive n sufficiently large.*

As a consequence of Proposition 1, there is a least positive integer, n_0, such that (cf. (1.12))

$$\{z_{k,n}\}_{k=1}^{n} \bigcap \mathcal{D}_n \neq \emptyset \text{ for all positive integers } n > n_0, \tag{1.14}$$

i.e., at least one zero of $s_n(nz)$ lies in $\mathcal{D}_n$ for every $n > n_0$. By direct calculation of the zeros of $s_n(nz)$, it appears that

$$n_0 = 96, \tag{1.15}$$

and also that

$$\{z_{k,n}\}_{k=1}^{n} \bigcap \mathcal{D}_n = \emptyset \quad (n = 1, 2, \cdots, n_0). \tag{1.16}$$

The *size* of $n_0 = 96$ is somewhat surprising. Because n_0 is so large, it was necessary to calculate the zeros of $s_n(nz)$ with great precision, and for this, Richard Brent's MP package was used with 120 significant digits.

As a consequence of Proposition 1, it is natural to ask if there is a simple *modification*, say $\hat{\mathcal{D}}_n$, of the definition of the closed set $\mathcal{D}_n$ of (1.12) which would have all the zeros $\{z_{k,n}\}_{k=1}^{n}$ *outside* $\hat{\mathcal{D}}_n$ for *all* $n \geq 1$. To give an affirmative answer to this question, we define, for each $n = 1, 2, \cdots$, the arc

$$\hat{D}_n := \Big\{ z \in \mathbb{C} : |ze^{1-z}|^n = \tau_n \sqrt{2\pi n} \left| \frac{1 - Re\ z}{z} \right|, \quad |z| \leq 1, \text{ and } |\arg z| \geq \cos^{-1}\left(\frac{n-2}{n}\right) \Big\}, \tag{1.17}$$

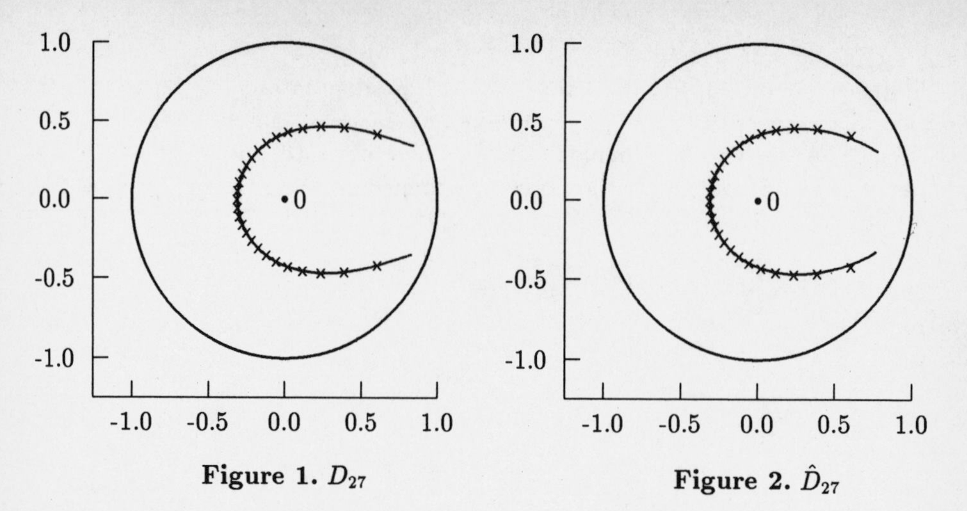

Figure 1. D_{27} **Figure 2.** $\hat{D}_{27}$

and its associated closed star-shaped (with respect to $z = 0$) set

$$(1.18)\quad \hat{\mathcal{D}}_n := \left\{ z \in \mathbb{C} : |ze^{1-z}|^n \leq \tau_n \sqrt{2\pi n} \left| \frac{1 - Re\ z}{z} \right|, \quad |z| \leq 1, \text{ and } |\arg z| \geq \cos^{-1}\left(\frac{n-2}{n}\right) \right\}.$$

Unfortunately, this modification does *not* preserve the accuracy of (1.10). Our main result (which will be sketched in §2) is

Theorem 2 *With the definition of the set* $\hat{\mathcal{D}}_n$ *of* (1.18), *then*

$$(1.19)\quad \{z_{k,n}\}_{k=1}^{n} \bigcap \hat{\mathcal{D}}_n = \emptyset \qquad (n = 1, 2, \cdots),$$

and, with the definition of (1.5),

$$(1.20)\quad \text{dist}\ [\{z_{k,n}\}_{k=1}^{n} \backslash C_\delta; \hat{\mathcal{D}}_n] = O\left(\frac{1}{n}\right) \qquad (n \to \infty),$$

for any fixed δ *with* $0 < \delta \leq 1$.

We remark that the bound of (1.20) is *best possible*, as a function of n. We include here Figures 1 and 2 which respectively display the arcs D_{27} and $\hat{D}_{27}$, along with the zeros, $\{z_{k,27}\}_{k=1}^{27}$, of $s_{27}(27z)$. These zeros are denoted by $\times$'s on Figures 1 and 2. Figures 3 and 4 similarly display the arcs D_{49} and $\hat{D}_{49}$, along with the zeros $\{z_{k,49}\}_{k=1}^{49}$ of $s_{49}(49z)$.

2. Proof of Proposition 1

It is easy to verify (by differentiation) that

$$(2.1)\quad e^{-z}s_n(z) = 1 - \frac{1}{n!}\int_0^z \zeta^n e^{-\zeta} d\zeta \qquad (z \in \mathbb{C},\ n = 0, 1, \cdots),$$

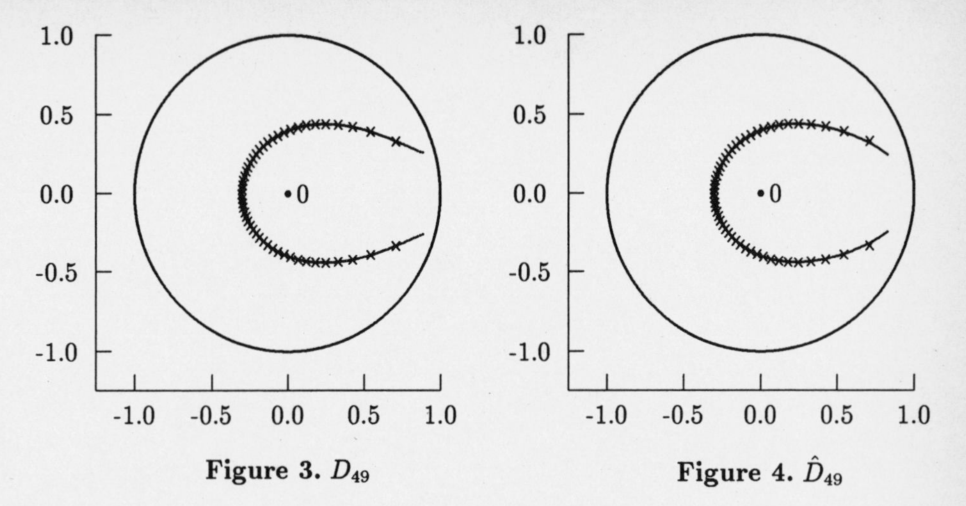

Figure 3. D_{49} **Figure 4.** $\hat{D}_{49}$

and replacing ζ and z, respectively, by $n\zeta$ and nz in(2.1) results in

$$e^{-nz}s_n(nz) = 1 - \frac{n^{n+1}}{n!}\int_0^z \zeta^n e^{-n\zeta} d\zeta. \tag{2.2}$$

With the definition of τ_n of (1.9), the above equation becomes

$$e^{-nz}s_n(nz) = 1 - \frac{\sqrt{n}}{\tau_n\sqrt{2\pi}}\int_0^z (\zeta e^{1-\zeta})^n d\zeta \qquad (z \in \mathbb{C},\ n = 0, 1, \cdots). \tag{2.3}$$

Now, in [3, eq. (2.14)], it is shown that

$$e^{-nz}s_n(nz) = 1 - \frac{z(ze^{1-z})^n}{\tau_n\sqrt{2\pi n}(1-z)}\left\{1 - \frac{1}{(n+1)(1-z)^2} + O\left(\frac{1}{n^2}\right)\right\}, \tag{2.4}$$

uniformly on any compact subset Ω of $\bar{\Delta}\backslash\{1\}$. On fixing Ω, then for any zero, $z_{k,n}$, of $s_n(nz)$ in Ω, we evidently have from (2.4) that

$$\frac{z_{k,n}(z_{k,n}e^{1-z_{k,n}})^n}{\tau_n\sqrt{2\pi n}(1-z_{k,n})}\left\{1 - \frac{1}{(n+1)(1-z_{k,n})^2} + O\left(\frac{1}{n^2}\right)\right\} = 1, \tag{2.5}$$

so that

$$\frac{|z_{k,n}(z_{k,n}e^{1-z_{k,n}})^n|}{\tau_n\sqrt{2\pi n}\cdot|1-z_{k,n}|}\left\{\left|1 - \frac{1}{(n+1)(1-z_{k,n})^2} + O\left(\frac{1}{n^2}\right)\right|\right\} = 1. \tag{2.6}$$

It is now clear that the arc D_n of (1.8), is just the approximation of (2.6), with a continuous variable z, when the quantity in braces in (2.6) is replaced by unity, i.e.,

$$\lambda_n(z) := \frac{|z(ze^{1-z})^n|}{\tau_n\sqrt{2\pi n}|1-z|}. \tag{2.7}$$

It is further evident from (1.12) that

(2.8) a zero $z_{k,n}$ of $s_n(nz)$ lies in $\mathcal{D}_n$ iff $\lambda_n(z_{k,n}) \le 1$.

We now examine the particular zero $z_{1,n}$ of $s_n(nz)$ which has smallest argument (cf. (1.13)). As discussed in [3, eq. (2.3)], we can write

$$z_{1,n} = 1 + \sqrt{\frac{2}{n}}(t_1 + o(1)) \qquad (n \to \infty), \tag{2.9}$$

where t_1 is the zero of $\mathrm{erfc}(w) := \frac{2}{\sqrt{\pi}}\int_w^\infty e^{-t^2} dt$ in the upper half-plane (i.e., $\mathrm{Im}\, t_1 > 0$) which is closest to the origin, $w = 0$, and it is known numerically (cf. Fettis, Caslin, and Cramer [4]) that

$$t_1 = -1.354810\cdots + i1.991467\cdots . \tag{2.10}$$

On evaluating $\lambda_n(z_{1,n})$ from (2.7), (2.9), and (2.10), it can be verified that

$$\lim_{n\to\infty} \lambda_n(z_{1,n}) = 0.985964\cdots < 1. \tag{2.11}$$

Thus, with (2.8), $z_{1,n}$ is contained in $\mathcal{D}_n$ for all positive n sufficiently large, which establishes Proposition 1. ■

We remark that it is because the constant, $0.985964\cdots$, of (2.11) is so close to unity, that it is *difficult* to see, graphically, that there are zeros of $s_n(nz)$ which lie interior to $\mathcal{D}_n$, for all n sufficiently large.

3. Proof of Theorem 2

We consider the integral (cf. (2.3))

$$I_n(z) := \int_0^z (\zeta e^{1-\zeta})^n d\zeta \qquad (z \in \mathrm{C},\ n = 0, 1, \cdots), \tag{3.1}$$

and, with $z = re^{i\theta}$, we choose the line segment $\zeta = \rho e^{i\theta} (0 \le \rho \le r)$ for the path of integration in (3.1). Then,

$$|I_n(z)| \le \int_0^r (\rho e^{1-\rho\cos\theta})^n d\rho =: J_n(r;\theta). \tag{3.2}$$

For $\theta = \pm\pi/2$, we see that $J_n(r; \pm\pi/2)$ can be expressed as

$$J_n(r; \pm\pi/2) = \frac{r(re^{1-r\cos\theta})^n}{n+1} < \frac{r(re^{1-r\cos\theta})^n}{n(1 - r\cos\theta)}. \tag{3.3i}$$

When $\cos\theta < 0$, $J_n(r;\theta)$ can be expressed as

$$J_n(r;\theta) = \frac{1}{|\cos\theta|^{n+1}} \int_0^{r|\cos\theta| e^{1+r|\cos\theta|}} v^{n-1} \frac{u(v)}{1 + u(v)} dv \qquad (v := ue^{1+u}).$$

Because $u/(1 + u)$ is strictly increasing, it can be verified that

$$J_n(r;\theta) < \frac{r(re^{1-r\cos\theta})^n}{n(1 - r\cos\theta)} \qquad (0 < r < 1, \text{ and } \cos\theta < 0). \tag{3.3ii}$$

But this same derivation also shows that the above holds for all $0 < r < 1$ and $\cos\theta > 0$. Thus, with (3.2) and (3.3), we have

$$|I_n(z)| < \frac{|z||ze^{1-z}|^n}{n(1-\text{Re } z)} \quad (0 < |z| < 1, \; n = 1, 2, \cdots), \tag{3.4}$$

and from (2.3), we further have that

$$\frac{\sqrt{n}}{\tau_n\sqrt{2\pi}} I_n(z) = 1 - e^{-nz} s_n(nz). \tag{3.5}$$

Thus, if $\{z_{k,n}\}_{k=1}^n$ denotes the set of zeros of $s_n(nz)$, then $\frac{\sqrt{n}}{\tau_n\sqrt{2\pi}} I_n(z_{k,n}) = 1$, which implies from (3.4) that

$$\frac{|z_{k,n}||z_{k,n}e^{1-z_{k,n}}|^n}{\tau_n\sqrt{2\pi n}(1-\text{Re } z_{k,n})} > 1. \tag{3.6}$$

From (1.18), this means that all zeros $\{z_{k,n}\}_{k=1}^n$ of $s_n(nz)$ lie *outside* the set $\hat{\mathcal{D}}_n$, for all $n \geq 1$, which is the desired result of (1.19) of Theorem 2.

The remainder of Theorem 2, to establish (1.20), now similarly follows, as in the proof given in [3, Theorem 4], by expressing a zero, $z_{k,n}$, of $s_n(nz)$, as $\hat{z} + \delta$, where $\hat{z}$ is a suitable boundary point of $\hat{\mathcal{D}}_n$, and where δ is assumed small. This argument also shows that the result of (1.20) is best possible. ∎

References

[1] N. Anderson, E.B. Saff, and R.S. Varga, *On the Eneström-Kakeya Theorem and its sharpness*, Linear Algebra Appl. **28** (1979), 5-16.

[2] J.D. Buckholtz, *A characterization of the exponential series*, Amer. Math. Monthly **73**, Part II (1966), 121-123.

[3] A.J. Carpenter, R.S. Varga, and J. Waldvogel, *Asymptotics for the zeros of the partial sums of e^z. I.*, Rocky Mount. J. of Math. (to appear).

[4] H.E. Fettis, J.C. Caslin, and K.R. Cramer, *Complex zeros of the error function and of the complementary error function*, Math. Comp. **27** (1973), 401-404.

[5] E.B. Saff and R.S. Varga, *On the zeros and poles of Padé approximants to e^z*, Numer. Math. **25** (1975), 1-14.

[6] E.B. Saff and R.S. Varga, *Zero-free parabolic regions for sequences of polynomials*, SIAM J. Math. Anal. **7** (1976), 344-357.

[7] G. Szegö, *Über eine Eigenschaft der Exponentialreihe*, Sitzungsber. Berl. Math. Ges. **23** (1924), 50-64.

Received: February 28, 1990

Lectures presented during the conference

R. A. Askey, Madison, USA
"Polynomial inequalities".

R.W. Barnard, Texas Tech, USA
"On two conjectures in geometric function theory".

H.-P. Blatt, Eichstätt, FRG
"Erdös-Turán Theorems on Jordan curves and arcs".

P. Borwein, Halifax, Canada
"A remarkable cube mean iteration".

A. Córdova, Würzburg, FRG
"Maximal range problems for polynomials".

C. FitzGerald, San Diego, USA
"Slit Mapping problems with no corresponding extremal problem".

R. Fournier, Montréal, Canada
Starlike univalent functions bounded on a diameter".

R. Freund, Würzburg, FRG
"A constrained Chebyshev approximation problem for ellipses".

W.H.J. Fuchs, Cornell, USA
"On a conjecture of Fischer and Michelli".

W.K. Hayman, York, UK
"A functional equation arising from the mortality tables".

D. Hough, Zürich, Switzerland
"A Symm-Jacobi collocation method for numerical conformal mapping"

J.A. Hummel, Maryland, USA
"Numerical solutions of the Schiffer differential equation".

L. Jacobsen, Trondheim, Norway
"Orthogonal polynomials, chain sequences, three-term recurrence relations and continued fractions".

W.B. Jones, Boulder, USA
"Zeros of Szegö polynomials associated with Wiener-Levinson prediction".

A. Lewis, Halifax, Canada
"On the convergence of moment problems".

D. Mejía, Medellín, Colombia
"Hyperbolic geometry in spherically k-convex regions".

D. Minda, Cincinatti, USA
"An application of hyperbolic geometry".

P.D. Miletta, Santiago, Chile
"Approximation of special functions to delay differential equations".

P. Nevai, Ohio State, USA
"Computational aspects of orthogonal polynomials".

O. Orellana, Valparaíso, Chile
"On the point vortex method and some applications to problems in aerodynamics".

N. Papamichael, Brunel, UK
"A domain decomposition method for conformal mapping onto a rectangle".

C. Pommerenke, TU Berlin, FRG
"Conformal mapping and the computation of bad curves".

B. Rodin, San Diego, USA
"Circle packing and conformal mapping".

F. Rønning, Trondheim, Norway
"A result about the sections of univalent functions".

St. Ruscheweyh, Valparaíso, Chile, and Würzburg, FRG
"Convexity preserving operators".

A. Ruttan, Kent State, USA
"Optimal successive overrelaxation iterative methods for p-cyclic matrices".

E.B. Saff, Tampa, USA
"Distribution of extreme points on best complex polynomial approximation".

L. Salinas, Valparaíso, Chile
"On abstract conjugation and some engineering applications".

G. Schober, Bloomington, USA
"Planar harmonic mappings".

D.F. Shea, Madison, USA
"An extremal property of entire functions with positive zeros".

F. Stenger, Salt Lake City, USA
"Explicit exponential and rational approximation of continuous functions on $\mathbf{R}$".

H.J. Stetter, TH Vienna, Austria
"Numerical inversion of multivariate polynomial systems".

T.J. Suffridge, Lexington, USA
"On nonvanishing H^p functions".

W.J. Thron, Boulder, USA
"Consequences of separate convergence of continued fractions".

R.S. Varga, Kent State, USA
"Recent results on the Riemann Hypothesis".

R.A. Zalik, Auburn, USA
"On the nonlinear Jeffcott equations".

D. Zwick, Vermont, USA
"Recent progress on best harmonic and subharmonic approximation".

Other Chilean participants

J. Almanza, Concepción
J. Bestagno, Concepción
H. Burgos, Temuco
V. Gruenberg, Valparaíso
S. Martínez, Concepción
W. Moscoso, Temuco
F. Novoa, Concepción
H. Pinto, Valparaíso
V. Valderrama, Punta Arenas
J.C. Vega, Concepción
V. Vargas, Temuco

LECTURE NOTES IN MATHEMATICS

Edited by A. Dold, B. Eckmann and F. Takens

Some general remarks on the publication of proceedings of congresses and symposia

Lecture Notes aim to report new developments - quickly, informally and at a high level. The following describes criteria and procedures which apply to proceedings volumes. The editors of a volume are strongly advised to inform contributors about these points at an early stage.

§1. One (or more) expert participant(s) of the meeting should act as the responsible editor(s) of the proceedings. They select the papers which are suitable (cf. §§ 2, 3) for inclusion in the proceedings, and have them individually refereed (as for a journal). It should not be assumed that the published proceedings must reflect conference events faithfully and in their entirety. Contributions to the meeting which are not included in the proceedings can be listed by title. The series editors will normally not interfere with the editing of a particular proceedings volume - except in fairly obvious cases, or on technical matters, such as described in §§ 2, 3. The names of the responsible editors appear on the title page of the volume.

§2. The proceedings should be reasonably homogeneous (concerned with a limited area). For instance, the proceedings of a congress on "Analysis" or "Mathematics in Wonderland" would normally not be sufficiently homogeneous.

One or two longer survey articles on recent developments in the field are often very useful additions to such proceedings - even if they do not correspond to actual lectures at the congress. An extensive introduction on the subject of the congress would be desirable.

§3. The contributions should be of a high mathematical standard and of current interest. Research articles should present new material and not duplicate other papers already published or due to be published. They should contain sufficient information and motivation and they should present proofs, or at least outlines of such, in sufficient detail to enable an expert to complete them. Thus resumes and mere announcements of papers appearing elsewhere cannot be included, although more detailed versions of a contribution may well be published in other places later.

Contributions in numerical mathematics may be acceptable without formal theorems resp. proofs if they present new algorithms solving problems (previously unsolved or less well solved) or develop innovative qualitative methods, not yet amenable to a more formal treatment. .

Surveys, if included, should cover a sufficiently broad topic, and should in general not simply review the author's own recent research. In the case of such surveys, exceptionally, proofs of results may not be necessary.

§4. "Mathematical Reviews" and "Zentralblatt für Mathematik" recommend that papers in proceedings volumes carry an explicit statement that they are in final form and that no similar paper has been or is being submitted elsewhere, if these papers are to be considered for a review. Normally, papers that satisfy the criteria of the Lecture Notes in Mathematics series also satisfy this requirement, but we strongly recommend that the contributing authors be asked to give this guarantee explicitly at the beginning or end of their paper. There will occasionally be cases where this does not apply but where, for special reasons, the paper is still acceptable for LNM.

§5. Proceedings should appear soon after the meeeting. The publisher should, therefore, receive the complete manuscript (preferably in duplicate) within nine months of the date of the meeting at the latest.

§6. Plans or proposals for proceedings volumes should be sent to one of the editors of the series or to Springer-Verlag Heidelberg. They should give sufficient information on the conference or symposium, and on the proposed proceedings. In particular, they should contain a list of the expected contributions with their prospective length. Abstracts or early versions (drafts) of some of the contributions are helpful.

§7. Lecture Notes are printed by photo-offset from camera-ready typed copy provided by the editors. For this purpose Springer-Verlag provides editors with technical instructions for the preparation of manuscripts and these should be distributed to all contributing authors. Springer-Verlag can also, on request, supply stationery on which the prescribed typing area is outlined. Some homogeneity in the presentation of the contributions is desirable.

Careful preparation of manuscripts will help keep production time short and ensure a satisfactory appearance of the finished book. The actual production of a Lecture Notes volume normally takes 6 -8 weeks.

Manuscripts should be at least 100 pages long. The final version should include a table of contents.

§8. Editors receive a total of 50 free copies of their volume for distribution to the contributing authors, but no royalties. (Unfortunately, no reprints of individual contributions can be supplied.) They are entitled to purchase further copies of their book for their personal use at a discount of 33.3 %, other Springer mathematics books at a discount of 20 % directly from Springer-Verlag. Contributing authors may purchase the volume in which their article appears at a discount of 33.3 %.

Commitment to publish is made by letter of intent rather than by signing a formal contract. Springer-Verlag secures the copyright for each volume.

<u>Addresses:</u>

Professor A. Dold, Mathematisches Institut, Universität Heidelberg, Im Neuenheimer Feld 288, 6900 Heidelberg, Federal Republic of Germany

Professor B. Eckmann, Mathematik, ETH-Zentrum
8092 Zürich, Switzerland

Prof. F. Takens, Mathematisch Instituut, Rijksuniversiteit Groningen, Postbus 800, 9700 AV Groningen, The Netherlands

Springer-Verlag, Mathematics Editorial, Tiergartenstr. 17,
6900 Heidelberg, Federal Republic of Germany, Tel.: (06221) 487-410

Springer-Verlag, Mathematics Editorial, 175, Fifth Avenue,
New York, New York 10010, USA, Tel.: (212) 460-1596

LECTURE NOTES

ESSENTIALS FOR THE PREPARATION
OF CAMERA-READY MANUSCRIPTS

The preparation of manuscripts which are to be reproduced by photo-offset require special care. Manuscripts which are submitted in technically unsuitable form will be returned to the author for retyping. There is normally no possibility of carrying out further corrections after a manuscript is given to production. Hence it is crucial that the following instructions be adhered to closely. If in doubt, please send us 1 - 2 sample pages for examination.

General. The characters must be uniformly black both within a single character and down the page. Original manuscripts are required: photocopies are acceptable only if they are sharp and without smudges.

On request, Springer-Verlag will supply special paper with the text area outlined. The standard TEXT AREA (OUTPUT SIZE if you are using a 14 point font) is 18 x 26.5 cm (7.5 x 11 inches). This will be scale-reduced to 75% in the printing process. If you are using computer typesetting, please see also the following page.

Make sure the TEXT AREA IS COMPLETELY FILLED. Set the margins so that they precisely match the outline and type right from the top to the bottom line. (Note that the page number will lie outside this area). Lines of text should not end more than three spaces inside or outside the right margin (see example on page 4).

Type on one side of the paper only.

Spacing and Headings (Monographs). Use ONE-AND-A-HALF line spacing in the text. Please leave sufficient space for the title to stand out clearly and do NOT use a new page for the beginning of subdivisons of chapters. Leave THREE LINES blank above and TWO below headings of such subdivisions.

Spacing and Headings (Proceedings). Use ONE-AND-A-HALF line spacing in the text. Do not use a new page for the beginning of subdivisons of a single paper. Leave THREE LINES blank above and TWO below headings of such subdivisions. Make sure headings of equal importance are in the same form.

The first page of each contribution should be prepared in the same way. The title should stand out clearly. We therefore recommend that the editor prepare a sample page and pass it on to the authors together with these instructions. Please take the following as an example. Begin heading 2 cm below upper edge of text area.

MATHEMATICAL STRUCTURE IN QUANTUM FIELD THEORY

John E. Robert
Mathematisches Institut, Universität Heidelberg
Im Neuenheimer Feld 288, D-6900 Heidelberg

Please leave THREE LINES blank below heading and address of the author, then continue with the actual text on the same page.

Footnotes. These should preferable be avoided. If necessary, type them in SINGLE LINE SPACING to finish exactly on the outline, and separate them from the preceding main text by a line.

Symbols. Anything which cannot be typed may be entered by hand in BLACK AND ONLY BLACK ink. (A fine-tipped rapidograph is suitable for this purpose; a good black ball-point will do, but a pencil will not). Do not draw straight lines by hand without a ruler (not even in fractions).

Literature References. These should be placed at the end of each paper or chapter, or at the end of the work, as desired. Type them with single line spacing and start each reference on a new line. Follow "Zentralblatt für Mathematik"/"Mathematical Reviews" for abbreviated titles of mathematical journals and "Bibliographic Guide for Editors and Authors (BGEA)" for chemical, biological, and physics journals. Please ensure that all references are COMPLETE and ACCURATE.

IMPORTANT

Pagination. For typescript, number pages in the upper right-hand corner in LIGHT BLUE OR GREEN PENCIL ONLY. The printers will insert the final page numbers. For computer type, you may insert page numbers (1 cm above outer edge of text area).

It is safer to number pages AFTER the text has been typed and corrected. Page 1 (Arabic) should be THE FIRST PAGE OF THE ACTUAL TEXT. The Roman pagination (table of contents, preface, abstract, acknowledgements, brief introductions, etc.) will be done by Springer-Verlag.

If including running heads, these should be aligned with the inside edge of the text area while the page number is aligned with the outside edge noting that right-hand pages are odd-numbered. Running heads and page numbers appear on the same line. Normally, the running head on the left-hand page is the chapter heading and that on the right-hand page is the section heading. Running heads should not be included in proceedings contributions unless this is being done consistently by all authors.

Corrections. When corrections have to be made, cut the new text to fit and paste it over the old. White correction fluid may also be used.

Never make corrections or insertions in the text by hand.

If the typescript has to be marked for any reason, e.g. for provisional page numbers or to mark corrections for the typist, this can be done VERY FAINTLY with BLUE or GREEN PENCIL but NO OTHER COLOR: these colors do not appear after reproduction.

COMPUTER-TYPESETTING. Further, to the above instructions, please note with respect to your printout that
- the characters should be sharp and sufficiently black;
- it is not strictly necessary to use Springer's special typing paper. Any white paper of reasonable quality is acceptable.

If you are using a significantly different font size, you should modify the output size correspondingly, keeping length to breadth ratio 1 : 0.68, so that scaling down to 10 point font size, yields a text area of 13.5 x 20 cm (5 3/8 x 8 in), e.g.

Differential equations.: use output size 13.5 x 20 cm.

Differential equations.: use output size 16 x 23.5 cm.

Differential equations.: use output size 18 x 26.5 cm.

Interline spacing: 5.5 mm base-to-base for 14 point characters (standard format of 18 x 26.5 cm).
If in any doubt, please send us 1 - 2 sample pages for examination. We will be glad to give advice.

Vol. 1259: F. Cano Torres, Desingularization Strategies for Three-Dimensional Vector Fields. IX, 189 pages. 1987.

Vol. 1260: N.H. Pavel, Nonlinear Evolution Operators and Semigroups. VI, 285 pages. 1987.

Vol. 1261: H. Abels, Finite Presentability of S-Arithmetic Groups. Compact Presentability of Solvable Groups. VI, 178 pages. 1987.

Vol. 1262: E. Hlawka (Hrsg.), Zahlentheoretische Analysis II. Seminar, 1984–86. V, 158 Seiten. 1987.

Vol. 1263: V.L. Hansen (Ed.), Differential Geometry. Proceedings, 1985. XI, 288 pages. 1987.

Vol. 1264: Wu Wen-tsün, Rational Homotopy Type. VIII, 219 pages. 1987.

Vol. 1265: W. Van Assche, Asymptotics for Orthogonal Polynomials. VI, 201 pages. 1987.

Vol. 1266: F. Ghione, C. Peskine, E. Sernesi (Eds.), Space Curves. Proceedings, 1985. VI, 272 pages. 1987.

Vol. 1267: J. Lindenstrauss, V.D. Milman (Eds.), Geometrical Aspects of Functional Analysis. Seminar. VII, 212 pages. 1987.

Vol. 1268: S.G. Krantz (Ed.), Complex Analysis. Seminar, 1986. VII, 195 pages. 1987.

Vol. 1269: M. Shiota, Nash Manifolds. VI, 223 pages. 1987.

Vol. 1270: C. Carasso, P.-A. Raviart, D. Serre (Eds.), Nonlinear Hyperbolic Problems. Proceedings, 1986. XV, 341 pages. 1987.

Vol. 1271: A.M. Cohen, W.H. Hesselink, W.L.J. van der Kallen, J.R. Strooker (Eds.), Algebraic Groups Utrecht 1986. Proceedings. XII, 284 pages. 1987.

Vol. 1272: M.S. Livšic, L.L. Waksman, Commuting Nonselfadjoint Operators in Hilbert Space. III, 115 pages. 1987.

Vol. 1273: G.-M. Greuel, G. Trautmann (Eds.), Singularities, Representation of Algebras, and Vector Bundles. Proceedings, 1985. XIV, 383 pages. 1987.

Vol. 1274: N. C. Phillips, Equivariant K-Theory and Freeness of Group Actions on C*-Algebras. VIII, 371 pages. 1987.

Vol. 1275: C.A. Berenstein (Ed.), Complex Analysis I. Proceedings, 1985–86. XV, 331 pages. 1987.

Vol. 1276: C.A. Berenstein (Ed.), Complex Analysis II. Proceedings, 1985–86. IX, 320 pages. 1987.

Vol. 1277: C.A. Berenstein (Ed.), Complex Analysis III. Proceedings, 1985–86. X, 350 pages. 1987.

Vol. 1278: S.S. Koh (Ed.), Invariant Theory. Proceedings, 1985. V, 102 pages. 1987.

Vol. 1279: D. Ieşan, Saint-Venant's Problem. VIII, 162 Seiten. 1987.

Vol. 1280: E. Neher, Jordan Triple Systems by the Grid Approach. XII, 193 pages. 1987.

Vol. 1281: O.H. Kegel, F. Menegazzo, G. Zacher (Eds.), Group Theory. Proceedings, 1986. VII, 179 pages. 1987.

Vol. 1282: D.E. Handelman, Positive Polynomials, Convex Integral Polytopes, and a Random Walk Problem. XI, 136 pages. 1987.

Vol. 1283: S. Mardešić, J. Segal (Eds.), Geometric Topology and Shape Theory. Proceedings, 1986. V, 261 pages. 1987.

Vol. 1284: B.H. Matzat, Konstruktive Galoistheorie. X, 286 pages. 1987.

Vol. 1285: I.W. Knowles, Y. Saitō (Eds.), Differential Equations and Mathematical Physics. Proceedings, 1986. XVI, 499 pages. 1987.

Vol. 1286: H.R. Miller, D.C. Ravenel (Eds.), Algebraic Topology. Proceedings, 1986. VII, 341 pages. 1987.

Vol. 1287: E.B. Saff (Ed.), Approximation Theory, Tampa. Proceedings, 1985–1986. V, 228 pages. 1987.

Vol. 1288: Yu. L. Rodin, Generalized Analytic Functions on Riemann Surfaces. V, 128 pages, 1987.

Vol. 1289: Yu. I. Manin (Ed.), K-Theory, Arithmetic and Geometry. Seminar, 1984–1986. V, 399 pages. 1987.

Vol. 1290: G. Wüstholz (Ed.), Diophantine Approximation and Transcendence Theory. Seminar, 1985. V, 243 pages. 1987.

Vol. 1291: C. Mœglin, M.-F. Vignéras, J.-L. Waldspurger, Correspondances de Howe sur un Corps p-adique. VII, 163 pages. 1987

Vol. 1292: J.T. Baldwin (Ed.), Classification Theory. Proceedings, 1985. VI, 500 pages. 1987.

Vol. 1293: W. Ebeling, The Monodromy Groups of Isolated Singularities of Complete Intersections. XIV, 153 pages. 1987.

Vol. 1294: M. Queffélec, Substitution Dynamical Systems – Spectral Analysis. XIII, 240 pages. 1987.

Vol. 1295: P. Lelong, P. Dolbeault, H. Skoda (Réd.), Séminaire d'Analyse P. Lelong – P. Dolbeault – H. Skoda. Seminar, 1985/1986. VII, 283 pages. 1987.

Vol. 1296: M.-P. Malliavin (Ed.), Séminaire d'Algèbre Paul Dubreil et Marie-Paule Malliavin. Proceedings, 1986. IV, 324 pages. 1987.

Vol. 1297: Zhu Y.-l., Guo B.-y. (Eds.), Numerical Methods for Partial Differential Equations. Proceedings. XI, 244 pages. 1987.

Vol. 1298: J. Aguadé, R. Kane (Eds.), Algebraic Topology, Barcelona 1986. Proceedings. X, 255 pages. 1987.

Vol. 1299: S. Watanabe, Yu.V. Prokhorov (Eds.), Probability Theory and Mathematical Statistics. Proceedings, 1986. VIII, 589 pages. 1988.

Vol. 1300: G.B. Seligman, Constructions of Lie Algebras and their Modules. VI, 190 pages. 1988.

Vol. 1301: N. Schappacher, Periods of Hecke Characters. XV, 160 pages. 1988.

Vol. 1302: M. Cwikel, J. Peetre, Y. Sagher, H. Wallin (Eds.), Function Spaces and Applications. Proceedings, 1986. VI, 445 pages. 1988.

Vol. 1303: L. Accardi, W. von Waldenfels (Eds.), Quantum Probability and Applications III. Proceedings, 1987. VI, 373 pages. 1988.

Vol. 1304: F.Q. Gouvêa, Arithmetic of p-adic Modular Forms. VIII, 121 pages. 1988.

Vol. 1305: D.S. Lubinsky, E.B. Saff, Strong Asymptotics for Extremal Polynomials Associated with Weights on $\mathbb{R}$. VII, 153 pages. 1988.

Vol. 1306: S.S. Chern (Ed.), Partial Differential Equations. Proceedings, 1986. VI, 294 pages. 1988.

Vol. 1307: T. Murai, A Real Variable Method for the Cauchy Transform, and Analytic Capacity. VIII, 133 pages. 1988.

Vol. 1308: P. Imkeller, Two-Parameter Martingales and Their Quadratic Variation. IV, 177 pages. 1988.

Vol. 1309: B. Fiedler, Global Bifurcation of Periodic Solutions with Symmetry. VIII, 144 pages. 1988.

Vol. 1310: O.A. Laudal, G. Pfister, Local Moduli and Singularities. V, 117 pages. 1988.

Vol. 1311: A. Holme, R. Speiser (Eds.), Algebraic Geometry, Sundance 1986. Proceedings. VI, 320 pages. 1988.

Vol. 1312: N.A. Shirokov, Analytic Functions Smooth up to the Boundary. III, 213 pages. 1988.

Vol. 1313: F. Colonius, Optimal Periodic Control. VI, 177 pages. 1988.

Vol. 1314: A. Futaki, Kähler-Einstein Metrics and Integral Invariants. IV, 140 pages. 1988.

Vol. 1315: R.A. McCoy, I. Ntantu, Topological Properties of Spaces of Continuous Functions. IV, 124 pages. 1988.

Vol. 1316: H. Korezlioglu, A.S. Ustunel (Eds.), Stochastic Analysis and Related Topics. Proceedings, 1986. V, 371 pages. 1988.

Vol. 1317: J. Lindenstrauss, V.D. Milman (Eds.), Geometric Aspects of Functional Analysis. Seminar, 1986–87. VII, 289 pages. 1988.

Vol. 1318: Y. Felix (Ed.), Algebraic Topology – Rational Homotopy. Proceedings, 1986. VIII, 245 pages. 1988

Vol. 1319: M. Vuorinen, Conformal Geometry and Quasiregular Mappings. XIX, 209 pages. 1988.

Vol. 1380: H.P. Schlickewei, E. Wirsing (Eds.), Number Theory, Ulm 1987. Proceedings. V, 266 pages. 1989.

Vol. 1381: J.-O. Strömberg, A. Torchinsky. Weighted Hardy Spaces. V, 193 pages. 1989.

Vol. 1382: H. Reiter, Metaplectic Groups and Segal Algebras. XI, 128 pages. 1989.

Vol. 1383: D.V. Chudnovsky, G.V. Chudnovsky, H. Cohn, M.B. Nathanson (Eds.), Number Theory, New York 1985–88. Seminar. V, 256 pages. 1989.

Vol. 1384: J. Garcia-Cuerva (Ed.), Harmonic Analysis and Partial Differential Equations. Proceedings, 1987. VII, 213 pages. 1989.

Vol. 1385: A.M. Anile, Y. Choquet-Bruhat (Eds.), Relativistic Fluid Dynamics. Seminar, 1987. V, 308 pages. 1989.

Vol. 1386: A. Bellen, C.W. Gear, E. Russo (Eds.), Numerical Methods for Ordinary Differential Equations. Proceedings, 1987. VII, 136 pages. 1989.

Vol. 1387: M. Petković, Iterative Methods for Simultaneous Inclusion of Polynomial Zeros. X, 263 pages. 1989.

Vol. 1388: J. Shinoda, T.A. Slaman, T. Tugué (Eds.), Mathematical Logic and Applications. Proceedings, 1987. V, 223 pages. 1989.

Vol. 1000: Second Edition. H. Hopf, Differential Geometry in the Large. VII, 184 pages. 1989.

Vol. 1389: E. Ballico, C. Ciliberto (Eds.), Algebraic Curves and Projective Geometry. Proceedings, 1988. V, 288 pages. 1989.

Vol. 1390: G. Da Prato, L. Tubaro (Eds.), Stochastic Partial Differential Equations and Applications II. Proceedings, 1988. VI, 258 pages. 1989.

Vol. 1391: S. Cambanis, A. Weron (Eds.), Probability Theory on Vector Spaces IV. Proceedings, 1987. VIII, 424 pages. 1989.

Vol. 1392: R. Silhol, Real Algebraic Surfaces. X, 215 pages. 1989.

Vol. 1393: N. Bouleau, D. Feyel, F. Hirsch, G. Mokobodzki (Eds.), Séminaire de Théorie du Potentiel Paris, No. 9. Proceedings. VI, 265 pages. 1989.

Vol. 1394: T.L. Gill, W.W. Zachary (Eds.), Nonlinear Semigroups, Partial Differential Equations and Attractors. Proceedings, 1987. IX, 233 pages. 1989.

Vol. 1395: K. Alladi (Ed.), Number Theory, Madras 1987. Proceedings. VII, 234 pages. 1989.

Vol. 1396: L. Accardi, W. von Waldenfels (Eds.), Quantum Probability and Applications IV. Proceedings, 1987. VI, 355 pages. 1989.

Vol. 1397: P.R. Turner (Ed.), Numerical Analysis and Parallel Processing. Seminar, 1987. VI, 264 pages. 1989.

Vol. 1398: A.C. Kim, B.H. Neumann (Eds.), Groups – Korea 1988. Proceedings. V, 189 pages. 1989.

Vol. 1399: W.-P. Barth, H. Lange (Eds.), Arithmetic of Complex Manifolds. Proceedings, 1988. V, 171 pages. 1989.

Vol. 1400: U. Jannsen. Mixed Motives and Algebraic K-Theory. XIII, 246 pages. 1990.

Vol. 1401: J. Steprāns, S. Watson (Eds.), Set Theory and its Applications. Proceedings, 1987. V, 227 pages. 1989.

Vol. 1402: C. Carasso, P. Charrier, B. Hanouzet, J.-L. Joly (Eds.), Nonlinear Hyperbolic Problems. Proceedings, 1988. V, 249 pages. 1989.

Vol. 1403: B. Simeone (Ed.), Combinatorial Optimization. Seminar, 1986. V, 314 pages. 1989.

Vol. 1404: M.-P. Malliavin (Ed.), Séminaire d'Algèbre Paul Dubreil et Marie-Paul Malliavin. Proceedings, 1987 – 1988. IV, 410 pages. 1989.

Vol. 1405: S. Dolecki (Ed.), Optimization. Proceedings, 1988. V, 223 pages. 1989.

Vol. 1406: L. Jacobsen (Ed.), Analytic Theory of Continued Fractions III. Proceedings, 1988. VI, 142 pages. 1989.

Vol. 1407: W. Pohlers, Proof Theory. VI, 213 pages. 1989.

Vol. 1408: W. Lück, Transformation Groups and Algebraic K-Theory. XII, 443 pages. 1989.

Vol. 1409: E. Hairer, Ch. Lubich, M. Roche. The Numerical Solution of Differential-Algebraic Systems by Runge-Kutta Methods. VII, 139 pages. 1989.

Vol. 1410: F.J. Carreras, O. Gil-Medrano, A.M. Naveira (Eds.), Differential Geometry. Proceedings, 1988. V, 308 pages. 1989.

Vol. 1411: B. Jiang (Ed.), Topological Fixed Point Theory and Applications. Proceedings, 1988. VI, 203 pages. 1989.

Vol. 1412: V.V. Kalashnikov, V.M. Zolotarev (Eds.), Stability Problems for Stochastic Models. Proceedings, 1987. X, 380 pages. 1989.

Vol. 1413: S. Wright, Uniqueness of the Injective III₁ Factor. III, 108 pages. 1989.

Vol. 1414: E. Ramírez de Arellano (Ed.), Algebraic Geometry and Complex Analysis. Proceedings, 1987. VI, 180 pages. 1989.

Vol. 1415: M. Langevin, M. Waldschmidt (Eds.), Cinquante Ans de Polynômes. Fifty Years of Polynomials. Proceedings, 1988. IX, 235 pages. 1990.

Vol. 1416: C. Albert (Ed.), Géométrie Symplectique et Mécanique. Proceedings, 1988. V, 289 pages. 1990.

Vol. 1417: A.J. Sommese, A. Biancofiore, E.L. Livorni (Eds.). Algebraic Geometry. Proceedings, 1988. V, 320 pages. 1990.

Vol. 1418: M. Mimura, Homotopy Theory and Related Topics. Proceedings, 1988. V, 241 pages. 1990.

Vol. 1419: P.S. Bullen, P.Y. Lee, J.L. Mawhin, P. Muldowney, W.F. Pfeffer (Eds.), New Integrals. Proceedings, 1988. V, 202 pages. 1990.

Vol. 1420: M. Galbiati, A. Tognoli (Eds.), Real Analytic Geometry. Proceedings, 1988. IV, 366 pages. 1990.

Vol. 1421: H.A. Biagioni, A Nonlinear Theory of Generalized Functions. XII, 214 pages. 1990.

Vol. 1422: V. Villani (Ed.), Complex Geometry and Analysis. Proceedings, 1988. V, 109 pages. 1990.

Vol. 1423: S.O. Kochman, Stable Homotopy Groups of Spheres: A Computer-Assisted Approach. VIII, 330 pages. 1990.

Vol. 1424: F.E. Burstall, J.H. Rawnsley, Twistor Theory for Riemannian Symmetric Spaces. III, 112 pages. 1990.

Vol. 1425: R.A. Piccinini (Ed.), Groups of Self-Equivalences and Related Topics. Proceedings, 1988. V, 214 pages. 1990.

Vol. 1426: J. Azéma, P.A. Meyer, M. Yor (Eds.), Séminaire de Probabilités XXIV 1988/89. V, 490 pages. 1990.

Vol. 1428: K. Erdmann, Blocks of Tame Representation Type and Related Algebras. XV, 312 pages. 1990.

Vol. 1429: S. Homer, A. Nerode, R.A. Platek, G.E. Sacks, A. Scedrov, Logic and Computer Science. Seminar, 1988. Editor: P. Odifreddi. V, 162 pages. 1990.

Vol. 1430: W. Bruns, A. Simis (Eds.), Commutative Algebra. Proceedings, 1988. V, 160 pages. 1990.

Vol. 1431: J.G. Heywood, K. Masuda, R. Rautmann, V.A. Solonnikov (Eds.), The Navier-Stokes Equations – Theory and Numerical Methods. Proceedings, 1988. VII, 238 pages. 1990.

Vol. 1432: K. Ambos-Spies, G.H. Müller, G.E. Sacks (Eds.), Recursion Theory Week. Proceedings, 1989. VI, 393 pages. 1990.

Vol. 1433: S. Lang, W. Cherry, Topics in Nevanlinna Theory. III, 174 pages. 1990.

Vol. 1435: St. Ruscheweyh, E.B. Saff, L.C. Salinas, R.S. Varga (Eds.), Computational Methods and Function Theory. Proceedings, 1989. VI, 211 pages. 1990.